BABYLON
TO
BAKU

by Zayn Bilkadi

GRIFFIN GOLD ARMLET, FROM THE
OXUS TREASURE, NORTHERN BACTRIA 500 BC.
MASSIVE GRIFFIN HEADED ARMLET. THE RECESSED
PORTIONS ORIGINALLY CONTAINED COLOURED STONE
AND GLASS INLAYS, WHICH APPEAR TO HAVE BEEN
REMOVED FOR REUSE IN ANTIQUITY. ARMLETS WERE
AMONG THE ITEMS CONSIDERED GIFTS OF HONOUR AT
THE PERSIAN COURT.

ISBN 0 9528816 0 8

DESIGNED AND PRODUCED IN ENGLAND BY
BARRETT HOWE GROUP LIMITED,
WINDSOR.

F O R E W O R D

Babylon to Baku recounts the story of a naturally occurring resource that has been utilised in parts of the Middle East for approximately 40 000 years, when it was first used to fix handles to stone tools and to waterproof baskets.

Archaeologists have grown increasingly interested in this history. Excavations are throwing new light on ancient trade patterns, indicating that the ancient peoples of Northern Iraq, South West Iran and the Dead Sea were the first to use this resource and in most cases continue to do so to this day.

Evidence of the importance attached to this material by early civilisations can be concluded from an excavation at the Mesopotamian city of Nippur in Southern Iraq, a site dating to 3000 BC, where it has been uncovered more often than almost any other material.

Early practical uses ranged from lining water pipes, as an adhesive and for securing valuable coloured inlays onto furniture, musical instruments and architectural decoration. Contemporary documents, written on clay, detail its shipments and uses.

The British Museum has a particularly rich collection of objects and texts which illustrate some of these ancient uses and many of these are pictured in this book.

Dr. St. John Simpson,
Curator,
Department of Western Asiatic Antiquities,
The British Museum.

AROUND 3000 BC THE LAND OF SUMER IN
MESOPOTAMIA WAS A FERTILE PLACE WITH
IRRIGATED FIELDS AND HOME TO A
SOPHISTICATED CIVILISATION.

THE FIRST PEOPLES TO USE BITUMEN ON A LARGE
SCALE ARE KNOWN AS THE 'UBAIDS' IN REFERENCE
TO THE MOUND TELL al-UBAID NEAR THE ANCIENT
RUINS OF UR IN SOUTHERN IRAQ

THE
SILK ROADS
OF THE
ANCIENT
NEAR EAST

The lands featured in Babylon to Baku cover an extensive part of the Asian Continent.

Ancient trade routes which crossed Asia, known as the Silk Roads, became the channels through which early technologies were exchanged from the shores of the Mediterranean to the heart of China.

Mesopotamia in the South meaning 'land between the rivers' is located in the historic region between the rivers Tigris and Euphrates. Here man's first cities were built, flourishing civilisations with the first organised government. Here too, around 3000 BC, man learned to write - arguably one of the most significant developments in history.

Thousands of miles to the North East in remote villages of China and at approximately the same period of history, man first searched for water below the surface of the earth. What he actually discovered was to have far reaching consequences for mankind.

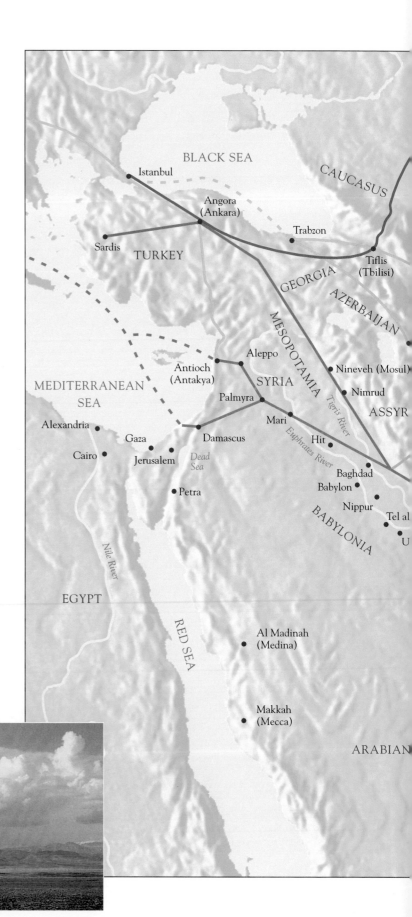

ASSYRIA

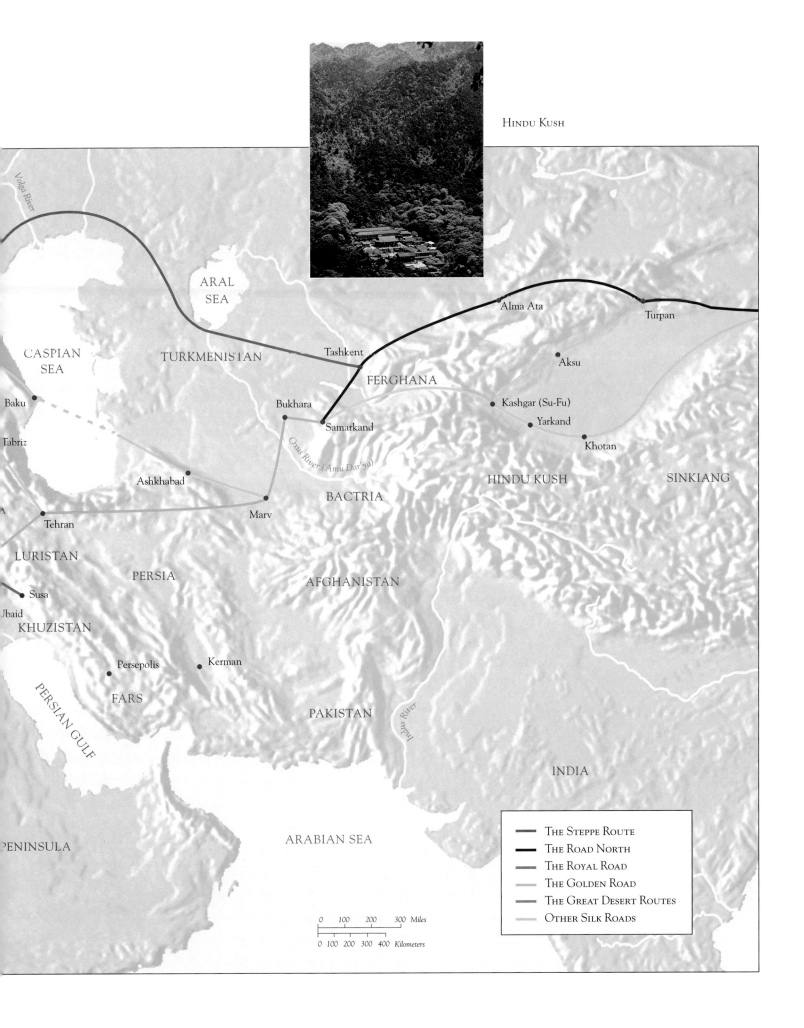

HINDU KUSH

Volga River

ARAL
SEA

CASPIAN
SEA

TURKMENISTAN

Alma Ata

Turpan

Baku

Tashkent

Aksu

FERGHANA

Tabriz

Bukhara

Kashgar (Su-Fu)

Samarkand

Yarkand

Oxus River (Amu Dar'ya)

Ashkhabad

Khotan

Tehran

Marv

BACTRIA

HINDU KUSH

SINKIANG

LURISTAN

PERSIA

Susa

Ubaid

AFGHANISTAN

KHUZISTAN

Persepolis

Kerman

FARS

PAKISTAN

Indus River

PERSIAN GULF

INDIA

PENINSULA

ARABIAN SEA

	THE STEPPE ROUTE
	THE ROAD NORTH
	THE ROYAL ROAD
	THE GOLDEN ROAD
	THE GREAT DESERT ROUTES
	OTHER SILK ROADS

0   100   200   300 Miles

0 100 200 300 400 Kilometers

7

CONTENTS

GLAZED TERRACOTTA TILE, FROM NIMRUD 860 BC.
AN ASSYRIAN KING WITH CUP IN ONE HAND AND BOW IN THE OTHER IS ACCOMPANIED BY HIS BODYGUARD AND ASSISTANT. PROBABLY PART OF A
MURAL DEPICTING THE KING AS A GREAT WARRIOR AND HUNTER, THIS TILE IS A VERY EARLY EXAMPLE OF COLOURED WALL DECORATION.

INTRODUCTION
BY THE AUTHOR

Today, there is one natural material which touches almost every facet of our lives; it assists us to travel long distances, it is an ingredient in many of our medicines, it is used in the manufacture of our clothes and in the microchips we build into our computers. In fact it is essential to our daily existence.

If mankind had not identified this material as a basic source of fuel long ago, then it is probable that the world would by now be witnessing global deforestation on a scale that is difficult to imagine.

Even the food we eat bears testimony to our dependence upon it, for without it as a source of the aids so essential to modern agriculture, the majority of the seven billion humans living today might well be doomed to famine.

Take it away and civilisation as we know it would surely and irrevocably change. We are truly living in an era in which we are wholly reliant upon it.

What is this natural material?

How, when and where did our reliance upon it take hold and begin to shape the course of human history?

Who were the people of the ancient civilisations who first recognised the importance of this, our most precious commodity?

Zayn Bilkadi

CHAPTER

ONE

SECRET

OF

BABYLONIA

*I*n about 450 BC the Greek historian
Herodotus, honoured among scholars as
the 'Father of History', left his homeland
in what is now South-Western Turkey
and embarked on a long journey to
explore Egypt and the Near East.

STATUE OF ASSYRIAN KING
ASSURBANIPAL II, FROM
NIMRUD 883 to 859 BC.
ASSURBANIPAL REBUILT THE CITY
OF KALHU NOW CALLED NIMRUD,
AS THE CAPITAL OF ASSYRIA.

(PREVIOUS PAGES) BABYLONIA - JOSEPH TURNER RA (1775-1851). BABYLONIA, MEANING 'DOOR OF GOD', IS THE SOUTHERN PORTION OF
MODERN-DAY IRAQ. IN ANTIQUITY, THE NORTHERN PART WAS REFERRED TO AS AKKAD, AND SOUTHERN PART AS SUMER.

THE HEAD OF A PERSIAN
GUARD, STONE RELIEF
FROM PERSEPOLIS
500 BC.

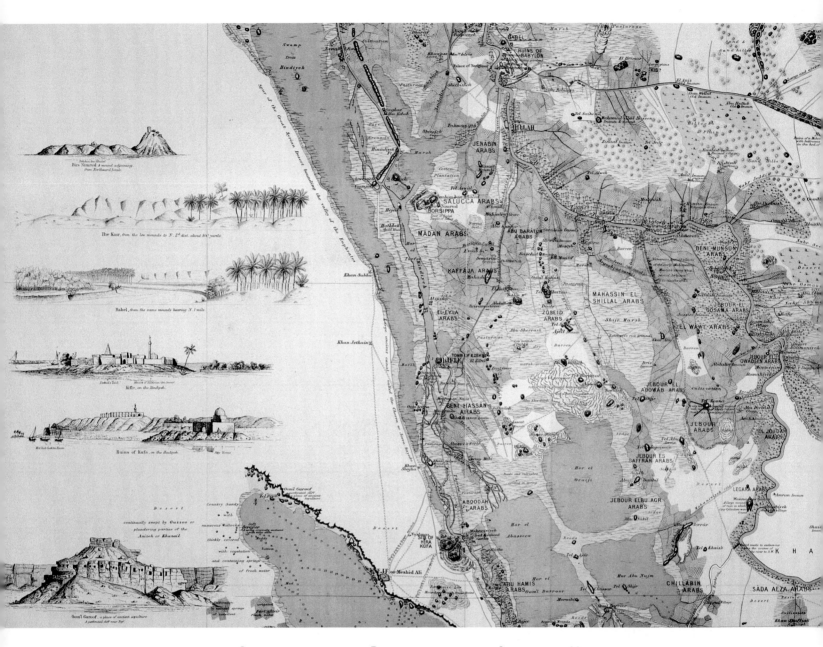

SURVEY MAP OF ANCIENT BABYLON, PUBLISHED IN LONDON IN 1885.

As he was travelling in the western province of Iran now known as Kuzistan near the Persian Gulf, he came across a most startling scene: a group of men working at a well out of which emerged not potable water, as one would expect, but lumps of asphalt mixed with salt-water and a strange oily substance that he could only call by its Persian name, *rhadinace*.

The workers, Herodotus noted, collected the oil by means of a long pole balanced horizontally on a pivot and at the end of which "*is attached (by a rope) half of a wineskin… and they dip with this, draw up the stuff, and pour it into a tank; from the tank it is drained off into another receptacle, and the three substances become separated: the asphalt and salt solidify at once.*"

CITY WALL OF BABYLON.
TWO MONUMENTS WERE ONCE CLASSIFIED AS WONDERS OF THE WORLD - THE HANGING GARDENS OF BABYLON, WHOSE
LOCATION REMAINS UNCERTAIN AND THE OTHER WAS THE CITY WALL, PERHAPS 18 KILOMETRES LONG, AND WHICH IS SAID TO
HAVE BEEN WIDE ENOUGH TO ALLOW FOR A FOUR HORSE CHARIOT TO TURN AT ITS TOP.

Golden chariot, from the Oxus treasure, Northern Bactria 500 bc.

The Persian word for the oil is rhadinace; it is very dark in colour, and has a strong smell."

Herodotus was obviously talking about crude oil, though he had no idea what this odiferous substance was good for, nor what its name was in Greek. So he left it at that - rhadinace - and never again mentioned this liquid in his writings. On the other hand, not far from this well, stood the border with Babylonia - now Southern Iraq, and in Babylonia at that time the people had a different, and decidedly more familiar, name for the dark oil.

They called it *naptu*, from the Akkadian verb *napatu* meaning 'to flare up'. So that in Akkadian, the language of ancient Iraq from about 2300 BC to 300 BC, *naptu* literally meant 'that which flares up', or 'that which catches fire', exactly for what the oily hydrocarbon is most noted.

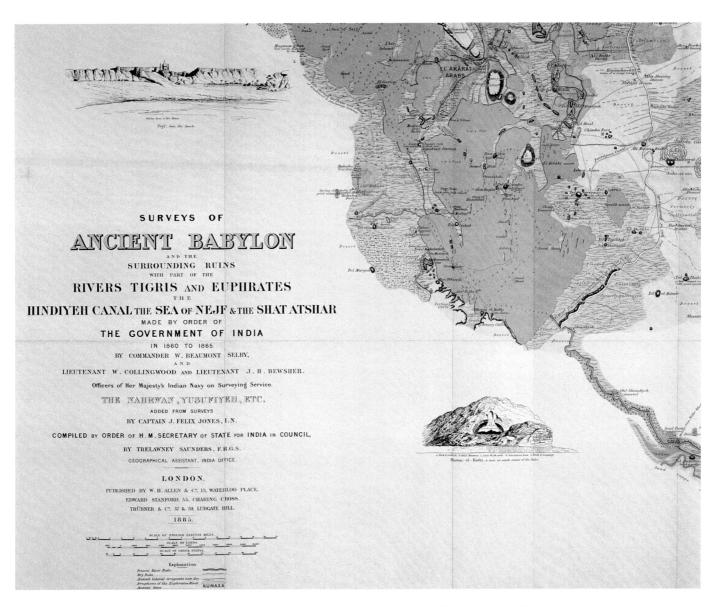

Survey map of ancient Babylon, published in London in 1885.

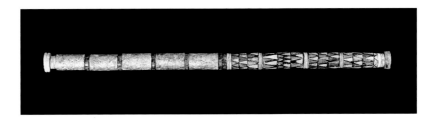

Mosaic sceptre, from Ur 2600 bc.
Inlaid with stone, shell and strips of gold foil using bitumen.
This may have been the handle of a fly whisk.

GOAT IN THE THICKET, FROM UR 2600 BC.
THIS DEPICTS A GOAT, PERCHED UP AGAINST A BUSH, A POSE OFTEN ADOPTED BY GOATS LOOKING FOR FOOD. THE TREE IS OF
GOLD LEAF AND THE GOAT HAS FACE AND LEGS OF GOLD LEAF. THE HORNS, EYES AND SHOULDER FLEECE ARE OF LAPIS LAZULI,
WHILE THE BODY FLEECE IS MADE OF WHITE SHELL, ALL CEMENTED WITH BITUMEN.

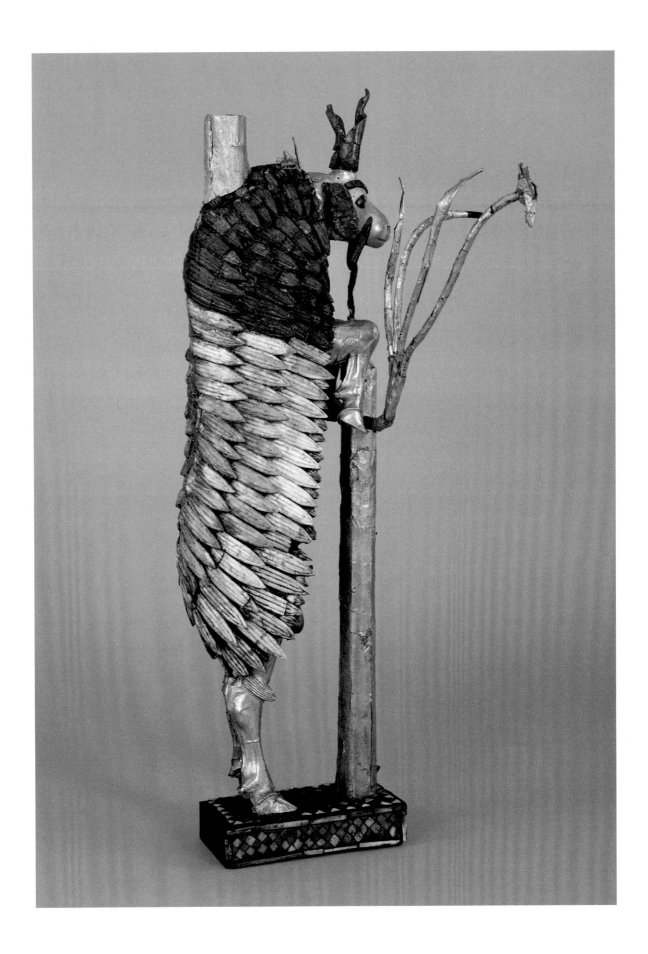

BUST OF ALEXANDER THE GREAT (356-323 BC). IN 336 BC, ALEXANDER EMBARKED ON A TERRITORIAL EXPANSION WHICH EXTENDED THE FRONTIERS OF THE GREEK EMPIRE TO EGYPT IN THE SOUTH AND INDIA IN THE EAST. HE DEFEATED THE PERSIAN ARMY IN 334 BC AND DIED IN BABYLON FROM FEVER AT THE AGE OF ONLY 33. HE WAS IMMORTALISED AS A GOD AFTER HIS DEATH AND WAS EVENTUALLY INTERRED AT ALEXANDRIA.

The flammability of *naptu* and its potential hazard had in fact been the subject of Babylonian writings since at least 2000 BC, long before Herodotus or any other Western writer had set foot in the Near East. Many cuneiform tablets dating from 2300-2000 BC have been recovered over the past century from the ruins of ancient Babylonian cities, which spoke of the ravages of oil fires and the sense of awe, if not helplessness, they instilled in the populace: "*if in a certain place of the land naptu oozes out, that country will walk in widowhood*" prophesied one tablet, and "*if a pit opens in fallow land*

and burning naptu appears, the land will be destroyed," added another from the same period. *Naptu* in fact is the root of the Arabic word for crude oil, *naft*, as well as the Hebrew *neft*, and, of course, the Greek *nafta* (written ναφτα) from which the modern English '*naphtha*' is derived. This explains why Herodotus, a man of extraordinary mastery of the Greek language, never used the word *nafta*.

Nafta did not enter the Greek language until a hundred years after his death, when the armies of Alexander the Great

invaded Babylonia and cross-cultural ties between the Greeks and Near Eastern people increased.

Because of his description of the well of Khuzistan, Herodotus earned himself another 'first' in history - he became the first man in the history of Western civilisation to report the sight of an oil well.

A NATURAL OIL SEEP.

(OPPOSITE PAGE) "*…Alexander the Great visited the town and watched its wells spew out the mysterious, magical oil*". *Alexander was one of the first Europeans to see natural oil seeps in the ancient Middle East.*

L'ENTRÉE D'ALEXANDRE
PRESENTÉE A

Par son tres humble, tres obeissant et tres fidele

DANS BABILONE
MONSEIGNEUR

Serviteur Seb. le Clerc Chevalier Romain.

LIONS HEAD MACE.
GOLD AND LAPIS.

But the well he described was neither unique nor man-made, as he probably thought. Rather, it was a natural seepage among countless others in the region. For in many areas of the Near East and Caucasus, especially in the landmass that stretches from the Persian Gulf to the Caspian Sea, oil and gas had been known since the beginning of human settlements in that part of the world. In fact some of these ominous-looking, and sometimes fiery, pools were the very signs that attracted the first European concessionaires to the region. In Baku, as in Iran, Iraq and Uzbekistan, the first oil wells sunk by the British and Russians in the 19th and early 20th centuries were invariably located near ancient seepages. For at this time, when the science of oil prospecting was still undeveloped, an oil seepage - or better still a burning gas well - was just about the only hint a drifter had that more substantial reserves might be hidden below the surface at that location.

Reports of oil and gas wells in such remote lands as Turkmenistan, Uzbekistan and even Afghanistan - two of which are now major producers of hydrocarbon - were recorded by many Greek and Roman writers between 300 BC and AD 100.

(PREVIOUS PAGES) 'ALEXANDER ENTERING BABYLON' - S. LE CLERC (1658). THE ARMY OF ALEXANDER THE GREAT PASSED THROUGH BACTRIA WHERE HISTORIANS AND TRAVELLERS HAD LONG NOTED BURNING OIL SEEPAGES. ALEXANDER ROUTINELY SENT INTERESTING NATURAL SPECIMENS BACK TO HIS TUTOR ARISTOTLE IN GREECE.

RECONSTRUCTION OF THE STATE APARTMENTS AT THE PALACE OF NIMRUD.
EACH WALL DEPICTS A SCENE PORTRAYING THE ASSYRIAN KING AS INVINCIBLE.

By far the most written-about oil wells in the ancient world were those in Iraq; mostly in the Southern half, traditionally known as Babylonia, and also in the Northern half, Assyria. There are some very good reasons why this was so, one of which is simply a question of numbers. Iraq had more oil seepages than any other land in antiquity. Sumerian, Babylonian, Greek and, much later, Arabic documents from the Middle Ages, spoke of oil and gas wells from the Upper Tigris River in the north of the

RELIEF SHOWING ALEXANDER'S TROOPS FIGHTING THE PERSIANS.
AFTER ALEXANDER'S DEATH IN 323 BC, HIS GENERALS DIVIDED HIS EMPIRE
INTO A SERIES OF INDEPENDENT KINGDOMS. THESE LANDS WERE
GRADUALLY 'HELLENISED'. 'HELLAS' MEANS GREECE, AND THE TERM
HELLENISTIC IS USED TO DESCRIBE THE PERIOD BETWEEN ALEXANDER'S
DEATH AND THE FALL OF EGYPT, THE LAST OF THESE INDEPENDENT
KINGDOMS, TO THE ADVANCING ROMAN EMPIRE.

country to the shores of Kuwait in the south. Such places as Zakho, Tuz Khormatu, Mosul, Kirkuk, Hit and Jabal Barma were all mentioned for their oil, as well as the now-lost sites of Magda and Kimas, said by the Babylonians of 2000 BC to lie in the mountains bordering Iran.

The medieval Arabs gave special names to these surface deposits. Al-Mas'udi, the famous Baghdad-born writer of the tenth century, known as the 'Herodotus of the Arabs,' coined the name *atam*, meaning 'pillar of fire' for the burning gas wells. As for the crude oil seepages, they had two names, depending on the viscosity of the oil.

Palace of Nimrud, restored.

Al-Jahiz, the ninth century essayist from Basra, called the naphtha spring a *naffata*, distinguishing it from the bitumen seepage, *qayyara* from the Arabic *qayr*, or bitumen. Two of the oldest and best-known oil sites were the seepages of Hit in Central Iraq, and Kirkuk in the North. Hit's original name in 3000 BC or earlier was '*Dul Dul*' which meant 'wells' in the Sumerian language, no doubt because of its oil. Kirkuk, on the other hand, is not a Sumerian name but Akkadian, dating to at least 1000 BC. In Akkadian it meant 'continuous murmur', due to the incessant humming of a burning gas well there since time immemorial.

CLAY TABLETS - WRITTEN IN 'CUNEIFORM' (MEANING WEDGE-SHAPED) ARE THE EARLIEST KNOWN EXAMPLES OF WRITING. HUGE NUMBERS OF THESE METICULOUSLY SCRIPTED RECORDS HAVE BEEN EXCAVATED FROM THE ANCIENT MIDDLE EAST AND DATE TO 3000 BC.

THE MANY THOUSANDS OF CUNEIFORM TABLETS RECOVERED FROM THE SANDS OF ANCIENT CITIES GIVE A FASCINATING AND DETAILED RECORD OF LIFE MORE THAN 5000 YEARS AGO. MERCHANT'S ACCOUNTS RECORD THE SALE OF 'KUPRU' AND 'ITTU' MEANING BITUMEN.
WRITING WAS ACHIEVED USING A REED STYLUS WHICH WAS SCRIBED INTO THE SOFT CLAY TABLET.
THE ART OF MAINTAINING SOFT CLAY, SUITABLE TO TAKE THESE DELICATE IMPRESSIONS IN HOT DESERT ENVIRONMENTS, HAS LONG SINCE BEEN LOST TO HISTORY.

(PREVIOUS PAGES) PALACE AT NIMRUD (NORTHERN IRAQ). BUILT BY KING ASSURBANIPAL II (860 BC) WHO INVITED 70 000 GUESTS TO THE OPENING CEREMONY.

To this day the name of that well, which in fact was the site of the first productive well in Iraq by the British Iraqi Oil Company, is Abu Kurkur (corrupted Abu Gurgur), meaning 'Father of Talk' in Arabic.

Aside from their sheer number, there are other more significant reasons why the oil deposits of ancient Iraq were the most celebrated of their kind. One of these had to do with the invention of writing, a phenomenon that triggered the first information revolution in history. The Sumerians, a people who lived in Southern Iraq from about 3000 BC to 2300 BC, developed writing before any other people in history; before the ancient Egyptians and long before the Chinese.

NINEVEH FROM THE NORTH EAST. NINEVEH (NOW MOSUL IN NORTHERN IRAQ) WAS THE ANCIENT CAPITAL OF THE ASSYRIAN EMPIRE, A COSMOPOLITAN CITY WITH SUPERB PALACES BUT INCREASINGLY VULNERABLE TO THREATS, PARTICULARLY FROM THE PROUD CITY OF BABYLON WHICH WAS RELUCTANT TO ACCEPT ASSYRIAN RULE. IT WAS CAPTURED AND SACKED BY BABYLON IN 612 BC.

31

SHRINE OF THE GOD HENDURSAG,
FROM UR 1750 BC.
WHEN DISCOVERED IT WAS REVEALED THAT
THE STATUE HAD BEEN BROKEN AND REPAIRED
USING BITUMEN IN ANCIENT TIMES. THE INSET
SHOWS THE STATUE AS SEEN TODAY.

The script they invented, called cuneiform, was written by pressing the tip of a reed stylus into a wet clay tablet. Not only did they invent this script, but together with their successors, the Babylonians, they practised it with zeal, leaving us meticulous records of practically every detail of what their world was about, including extensive information on oil. Archaeologists have in fact excavated tens of thousands of clay tablets from the ruins of ancient Iraq, including such Biblically- renowned cities as Babylon, Nineveh and Ur of the Chaldees. Some of these speak of oil wells, oil towns, oil spills, oil tankers, oil prices and even oil shortages. The picture that emerges from these writings is truly astounding: it is apparent that the Sumerians and Babylonians founded their civilisation on crude oil, making it into one of the most important pillars of their economy. Herodotus refers to *rhadinace* that oozed out of the well and noted that this was not in its pure state, but mixed with salt-water and asphalt. This, of course, is no surprise to anyone in the oil business today. Not only is salt-water almost always present in crude oil but, more importantly, crude oil comes in many grades.

In Akkadian, the heavy, semi-solid grades of crude oil were known by many names, the most common one being *kupru*. *Kupru* should not be a strange word in Western civilisation because it played a prominent, if not crucial, role in one of the most extraordinary epics in the Judeo-Christian faith - it was the very substance employed by the prophet Noah to waterproof his famous Ark and thus save humankind from the universal flood. In the Hebrew Bible, Noah used *koper*, later translated into the European languages as 'bitumen' or 'pitch.' In fact, the Hebrew *koper* and the Arabic *kafr*, both meaning bitumen rock, are direct descendants of the much older *kupru* of the Babylonians.

Though barely suitable as fuel, *kupru* for the ancient inhabitants of Iraq had an overwhelming commercial advantage over the lighter crude, *naptu*. By its very 'sticky' nature, the bitumen was eminently useful as a waterproof cement for securing or gluing together practically anything under the sun, from building materials, to tools, to art objects, to weapons. There is evidence, for example, that prehistoric hunters used the crude oil mastic to attach flint arrowheads to shafts, and that prehistoric farmers from the Neolithic Age harvested their wild crops with primitive stone sickles glued together with bitumen. A specimen of such a sickle excavated from the site of a Neolithic village called Umm el Tlel in Syria has recently been dated to around 38 000 BC.

SUMERIAN ROYAL HEAD-DRESS, FROM UR 2600 BC. TYPICAL JEWELLERY OF ANCIENT MESOPOTAMIA.

HEAD OF A WOMAN, FROM MARI 2000 BC.
BITUMEN WAS USED TO INSET HER EYES AND HIGHLIGHT
HER FACIAL DETAILS.

CARVED BITUMEN DRINKING CUP WITH GOAT
HANDLE, SUSA 2000 BC.

The first people to use bitumen on a large scale, however, belong to a culture known as the 'Ubaid' in reference to the mound Tell al-Ubaid near the ruins of ancient Ur, where artefacts were first unearthed some eighty years ago. These people settled in the Southern third of Iraq sometime around 6500 BC. There they practised crude irrigation farming, cultivated the date palm, and raised herds of domesticated cattle. Over the centuries they prospered and multiplied and built the foundations of most of the settlements along the Tigris and Euphrates Rivers where the great cities of the Sumerians and Babylonians were later to grow. Indeed, the names of most of these famed cities - Ur, Lagash, Eridu, Nippur and Kish are not Sumerian but Ubaidi and the first temples ever built in them were from the Ubaid period, which ended about 4000 BC.

SASSANIAN SILVER BOSS WITH THE FACE OF A LION, NORTHERN IRAN AD 400.

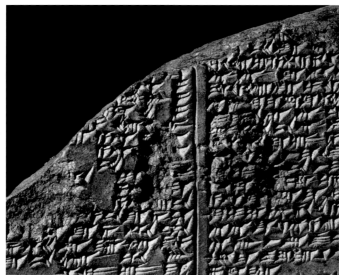

TABLET OF GILGAMESH 1635 BC.
ONE OF A SERIES OF CUNEIFORM SCRIPT
TABLETS RECOUNTING EPIC AND MYSTICAL
EVENTS, WRITTEN IN THE TIME OF KING
GILGAMESH OF ANCIENT MESOPOTAMIA.
THESE INCLUDE THE STORY OF A GREAT FLOOD
DURING WHICH A SHIP WAS BUILT AND
WATERPROOFED WITH PITCH, AND THE
SUBSEQUENT SAVING OF MANKIND - BEARING
A STRIKING RESEMBLANCE TO THE STORIES
TOLD IN THE BIBLE AND QUR'AN MORE THAN
3000 YEARS LATER.

(OPPOSITE PAGE) BABYLONIAN MAP OF THE WORLD 600 BC. THIS CELEBRATED MAP SHOWS THE WORLD AS A DISC SURROUNDED BY WATER.
BABYLON IS SHOWN IN THE CENTRE ON THE RIVER EUPHRATES WHICH FLOWS THROUGH THE MARSHES TO THE PERSIAN GULF AT THE
BASE OF THE TABLET. CIRCLES ARE USED TO INDICATE CITIES OR COUNTRIES. EIGHT OUTLYING REGIONS, TRIANGULAR IN SHAPE, ARE
THE HOME OF STRANGE OR LEGENDARY BEINGS. AT THE TOP IS WRITTEN 'WHERE THE SUN IS NOT SEEN' TO INDICATE NORTH.

Because stone and timber were scarce in their land, the Ubaids built shelters made from reeds which were abundant in the wetlands of the Shatt al-Arab. Their unique construction technique, dating back 5 000 to 6 000 years BC, is still followed today by the Marsh Arabs, called the Ma'dan, who inhabit that region. First, the reeds are assembled in long columns, which are tied firmly with bulrush fibre. These are then planted into the ground in two parallel rows with their tops bent inward and tied together in pairs to form the arches, and the completed frame is covered over with reed matting. As a final step, and for insulation purposes, they sometimes plastered the walls with a thick layer of bitumen. Not only was this material completely impervious to wind and moisture, but just as important, it was

BRACELET FROM THE JEWELLERY
COLLECTION OF LADY LAYARD.

maintenance-free, unlike a mud covering which often had to be repaired following flood waters or rainfall.

A fascinating twist in this construction technique was that it could just as easily be employed in building the hull of a reed ship upside down, without a single nail being employed. It is not surprising therefore, that by the fifth millennium BC, the Ubaids were already roaming their marshes in simple reed boats caulked with bitumen. In these primitive crafts, the first seafarers in history ventured into the high seas of the Persian Gulf to eventually reach Bahrain, and even Oman, farther South. In fact, signs of the Ubaid culture, especially its distinctive pottery style, have been found from Oman to the Mediterranean coastline of Syria.

(ABOVE) CORACLE RELIEF, FROM THE
PALACE OF NIMRUD 860 BC.
*"…by the fifth millennium BC, the Ubaids
were already roaming their marshes in
simple reed boats caulked with bitumen"*.

(OPPOSITE) CIRCULAR BOATS (CALLED
'GUFFAH') MADE OF RUSHES AND COATED
WITH BITUMEN ARE STILL USED ON THE
TIGRIS AND EUPHRATES TODAY, JUST AS
THEY WERE THOUSANDS OF YEARS AGO.

BOAT MOULDED IN BITUMEN,
FROM UR 2300 BC.
WHEN EXCAVATED FROM
A GRAVE, THE BOAT
WHICH IS 0.6M IN
LENGTH, CONTAINED
A CARGO OF SMALL
TERRACOTTA POTS EACH
HOLDING A PRECIOUS
SUBSTANCE - POSSIBLY FOR USE
IN THE AFTERLIFE OR TO WARD OFF
EVIL SPIRITS.

'ARAB CARAVAN' "...*On the shore, crews of women and children sprinkled the hunks of tarry oil with sand, stuffed them into leather bags and loaded them onto camels for the long journey across the Negev and Sinai*".

In the alluvial plain north of the Shatt al-Arab meanwhile, ancient builders had reached for another raw material (mud) to mould the first building bricks. At first, they simply left the clumps of mud in the sun to dry, and then used them to lay the foundations of mankind's first cities, first temples and first palaces. As the pace of civilisation accelerated however, these farmers-turned-bricklayers came to realise that sun-dried bricks by themselves were not nearly strong enough to support the larger structures, particularly the larger temples, that they now aspired to build for their gods. So, ingenious as they were, they took to adding chopped straw and bitumen to the clay before they dried it. The new bricks were thus made stronger and more water-resistant.

Pottery jars decorated with asphalt, from Ur 1900 bc.

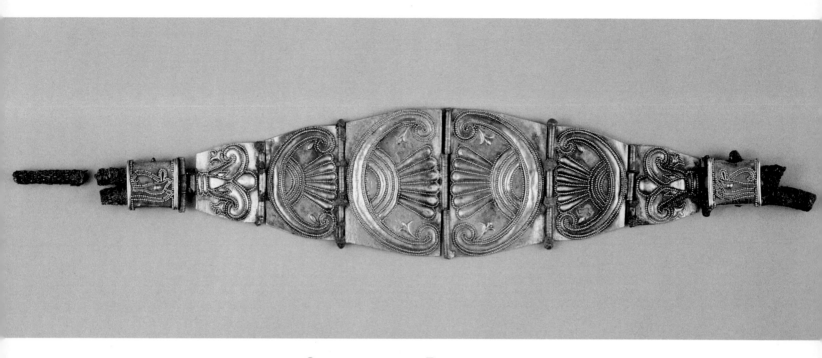

GOLD BRACELET, FROM THARROS 700 BC.
EIGHT HINGED SECTIONS EACH WITH INTRICATE GOLD SCROLLING.

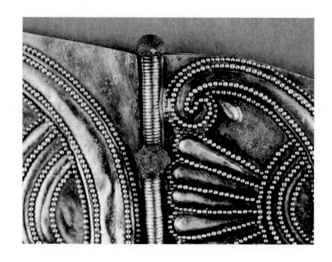

(PREVIOUS PAGES) RUINS OF THE TEMPLE OF PHILE.

By 4000 BC, in Mesopotamia, people had improved on the shape of the brick, making it flat on one side and convex on the other. More importantly, they had by now mastered the art of pottery production in the kiln and adapted this technique to brick-making on a large scale. As wood was scarce in their land, they must have used a mixture of oil and bitumen for fuel. This new brick had overwhelming advantages. In durability and strength it was far superior to the earlier brick. In addition, it had become porous - which meant that it could soak

BRICKWORK WITH BITUMEN MORTAR, FROM UR 2600 BC.

in bitumen like a sponge, and over time, become as hard as rock. For, unlike the ancient Egyptians and Romans who used limestone powder from which to make a cement paste, the ancient people of Iraq developed what proved to be an ideal complement to the porous brick: a petroleum mortar made from soft bitumen blended at times with small amounts of clay or chopped straw.

Because the brick was porous, setting it in this mortar would cause the bitumen to seep into the brick and eventually form a very strong joint - strong enough in fact, to last for several thousand years. Indications abound from clay tablets dating from 2500 BC, that the preparation of this special petroleum mortar had risen almost to a science complete with precise tabulations and mathematical formulae. Only in one other ancient civilisation, itself much influenced by Sumer, was such a petroleum cement employed on a large scale.

DRAGON'S HEAD BRICK, FROM ISHTAR GATE, BABYLON 580 BC.
IT HAS BEEN ESTIMATED THAT 15 MILLION BAKED BRICKS WERE USED
IN THE CONSTRUCTION OF BABYLON.

"…Labourers worked the oil wells, to obtain the oil; cameleers carried it and merchants sold it in the cities".

In Mohenjo Daro in today's Pakistan - an area known to the Babylonians as Meluhha - brick buildings contemporary with those of the Sumerians and Babylonians were cemented and waterproofed with bitumen.

Until about 2400 BC, many Sumerian cities were built with bitumen and the planoconvex bricks. Even the planoconvex bricks, however, were not ideally shaped; their edges, still too irregular, required too much bitumen or clay between the joints; and bitumen, for all its abundance, was never cheap.

RUINS OF AN ANCIENT ZIGGURAT 3000 BC.

(OPPOSITE) MASSIVE ZIGGURATS, OR
TEMPLE TOWERS, DOMINATED MANY
ANCIENT MESOPOTAMIAN CITIES.
THESE WERE CONSTRUCTED AND
DEDICATED TO SUCH GODS AS SHAMASH
(THE SUN GOD) AND SIN (THE MOON
GOD), THE MOST RENOWNED BEING THE
MYSTICAL 'TOWER OF BABEL'.

(OPPOSITE) A SACRIFICIAL ALTAR, ITS
SURFACE WATERPROOFED WITH BITUMEN.

At the peak of the monument-building boom in about 2300 BC, even the cheapest oil mastic fetched an already respectable half-ounce of silver per ton - a sum that was beyond the means of all but the wealthiest builders in the cities. And if the mastic needed further processing, especially melting or thinning, its price shot up rapidly, sometimes six or seven-fold, to make up for the added cost of the fuel, itself always in short supply.

Consequently, the planoconvex brick, the first 'oil guzzler' in the history of civilisation, had to be discarded. In its place appeared the 'cast' brick: instead of being moulded one at a time by hand, the bricks were now mass-produced by casting the mud in wooden frames of standard sizes. Trivial as it may now seem, this innovation was in fact a significant advance. Casting not only made for a cheaper and more regular shaped brick, but for a brick that required far less bitumen mortar as well because its edges were now perfectly flat.

MOSAIC
COLUMN, FROM
TELL AL-UBAID
2500 BC.
PROBABLY A TEMPLE
PILLAR, THIS WAS
ORIGINALLY OVER
THREE METRES HIGH
AND MADE FROM THE
TRUNK OF A PALM
TREE. THE DETAIL
SHOWS THE MOSAIC
OVERLAY IN RED
RIBBED STONE AND
MOTHER-OF-PEARL,
SET IN BITUMEN.

With this process in place, building costs declined, and the stage was at last set for the structures that were to dominate the landscape of the future Babylonian cities: lofty ziggurats, the huge defensive walls, bridges, dams and royal palaces.

In the beginning, the largest of these structures were the ziggurats which were massive temple towers erected at the centre of the city to lure the favours of the gods. With time, most cities had their own ziggurat, and many of these grew to colossal sizes requiring millions of bricks and thousands of tons of crude oil, both for firing the bricks and for serving as mortar. Among the most famous of these structures was the Tower of Babel, which took hundreds of years to complete.

One of the first Westerners to have seen this impressive monument happened to be, once again, the Greek historian Herodotus, who visited Babylon on his famous journey in about 450 BC. In approaching the city, which he described then as *"the most renowned and strongest"* of all the cities in the plain, what struck him first was the formidable brick wall surrounding it on all sides. It was a wall, he wrote, that was wide enough at its top to allow *"room for a four-horse chariot to turn"*.

Like most large structures in Babylonia, this outer wall had been built and rebuilt over the ages by successive kings, for it was Babylon's main defence against attacks from the outside world, and it would be hard to imagine the city surviving for as long as it did without it.

GOLD CUPS, FROM UR 2600 BC.

Yet what made the wall possible - and by the same token secured the city and the culture it stood for - was ultimately the crude oil used in its construction. Herodotus, in describing the huge moat that encircled the wall, paused to tell us how the wall was built:

"*As fast as they dug the moat, the soil which they got from the cutting was made into bricks, and when a sufficient number were completed they baked the bricks in kilns.*

Then they set to building, and began with bricking the borders of the moat, after which they proceeded to construct the wall itself, using throughout for their cement hot bitumen."

Between 1899 and 1913, a team of German archaeologists led by Robert Koldeway excavated debris of this imposing wall that once encircled the famous city. Koldeway painstakingly

exposed the foundations of two massive walls ten miles in circuit and each about twenty feet thick. Outside this double wall was the moat, each side of which was lined with a layer of burnt bricks, ten feet thick, set in bitumen. Given the dimensions of this immense structure, one can only conclude that the amount of crude transported to the site over the years to cement both the moat and the wall was enormous.

The Tower of Babel, as Herodotus saw it, stood in the heart of the city, and even though by that time it had been repeatedly assaulted and desecrated by the Persian conquerors, he still described it as a marvel of marvels: "*a tower of eight terraces stacked one above the other and resting on a square of solid masonry that covered ten acres of land*". Herodotus' words were supported by a late Babylonian inscription which was first uncovered in 1876 and is now in the British Museum.

The tablet, dated from about 350 BC, not only spoke of the ziggurat as a terraced tower rising almost 300 feet above the ground, but actually gave the exact dimensions of all but one of its terraces as follows:

"*Terrace 1: 300 feet on each side and 110 feet high.*

Terrace 2: 260 feet on each side and 60 feet high.

Terrace 3: 200 feet on each side and 20 feet high.

Terrace 4: 170 feet on each side and 20 feet high.

Terrace 5: 140 feet on each side and 20 feet high.

Terrace 6: Omitted.

Terrace 7: 80 feet long, 70 feet broad, 50 feet high.

'Ruins of a temple'.

"*This is the Temple of Bel,*" exclaimed the tablet, "*and in it was the statue of the God.*"

If we are to believe these figures, the total height of the tower would have been 280 feet, excluding the sixth terrace, which was probably omitted inadvertently. Such a monument would have been visible from sixty miles away in every direction; and an astounding ten million bricks, each one foot cube, would have been required to fill its first terrace alone. In restoring this ziggurat, the Babylonian King Nabopolassar (625-604 BC) boasted of endless "*torrents*" of bitumen (*kupru*) that he had shipped by boat for the project:

"*The lord Marduk commanded me concerning Etemenanki, the staged tower of Babylon, which in my time had become dilapidated and ruinous, that I should make its foundation secure in the bosom of the netherworld and make its summit like the heavens. I caused baked bricks to be made, as if they were rains from on high which are measureless, or great torrents. I caused streams of bitumen to be brought by the canal Arahtu ...*"

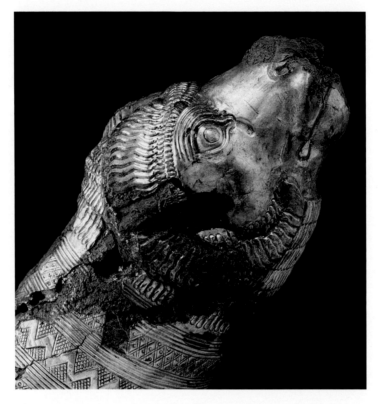

Silver ram's head, from Western Tehran 700 bc.
The silverwork has been formed over a base of asphalt.

Torrents of bitumen indeed would have been necessary to cement the tower together and to help bake the bricks needed for its outer skin. Although there is no way to estimate how much petroleum fuel was used to fire the kilns, we can make a rough guess from the geometry of the tower at how much bitumen mortar would have been required if the thickness of the outer walls was ten feet and the mortar between joints half an inch thick - several hundred tons at a minimum.

But where was all of this petroleum coming from? After all, it was not only Babylon that was built of baked bricks and bitumen, but many other cities and ziggurats in Southern Iraq as well. Here again we turn to Herodotus. For he too, after seeing the gigantic scale of the Babylonian monuments, wondered about the same thing. His answer points to a well known location outside of Babylon called Hit.

"*Eight days' journey from Babylon there is a city called Is on a smallish river of the same name, a tributary of the Euphrates, and in this river lumps of bitumen are found in great quantity. This was the source of supply for the bitumen used in building the wall of Babylon.*" It was also the source of bitumen for the building of the Tower of Babel, because Is at the time of Herodotus was the Greek name for the city of Hit.

(opposite) Straw boat relief, from the Palace of Nimrud 860 bc.
Stone relief showing soldiers crossing the marshes in a raft made from reed - these craft were often waterproofed with bitumen.

Of all the oil sites in ancient Iraq none surpassed in importance, in glory and in holiness the town of Hit - easily the oldest oil town in history and the original home of what may be considered the first petrochemical industry in the world. Over the centuries, the name of this 'oil Mecca' changed from Dul Dul, its oldest recorded name, to Idu and Ittu under the Sumerians, to Id under the early Babylonians and probably Ihi under the late Babylonians of 600 BC.

Its oil gushers apparently were so profuse and unrivalled that, after 2300 BC, its crude became generically known as *Iddu* - a word which literally meant 'from Id' - or 'the product from Id'.

Hit had an overwhelming geographic advantage as an oil town in Iraq. It sat almost exactly at the apex of the delta, making it the head of navigation on the Euphrates, and placing its oil within easy reach of every major city to the South,

from Babylon to Kish to Ur and even beyond, to the Persian Gulf countries of Bahrain and Oman as well. Its riches included hot and cold bitumen springs, light crude oil, sulphur mineral, and, to the delight of oracle and alchemist alike, several gas wells. These wells were called '*Usmeta*' in Assyrian - a word related to the later Arabic verb 'to hear' - because they hissed and murmured loudly as their gases jetted their way out through the narrow fissures in the ground.

GOLD DISK, FROM THE OXUS TREASURE, NORTHERN BACTRIA 500 BC.
ITEMS OF GOLD PROBABLY USED AS CLOTHING APPLIQUES.

THE ROYAL GAME OF UR 2600 BC.
EVIDENTLY ONE OF THE MOST POPULAR GAMES OF THE ANCIENT WORLD, GAME BOARDS OF THIS KIND WERE BEING MADE IN SYRIA BEFORE 3000 BC.
THE GAME WAS INTRODUCED INTO EGYPT ABOUT 1600 BC. CONSTRUCTED USING BITUMEN INLAY.

To the ancient kings of Mesopotamia, the wondrous Hit and its mysterious Usmeta became a destination of pilgrimage, a holy place where the muttering of the underworld gods could be heard.
The Assyrian King Tukulti Ninurta II (890-884 BC) wrote in his annals on the eve of one of his military campaigns: *"in front of Hit, by the bitumen (kupru) springs, the place of the Usmeta stones, in which the gods speak, I spent the night."*
As the economic pulse of Mesopotamia,

this land must have bustled in good times with hundreds of slave workers, contractors, shippers and accountants scrambling to fill orders from distant kings or distant countries. By some estimates, several hundred oil pits once dotted this town and the strip of land linking it to its two Southern neighbours, Ramadi and Abu Ghir.

Cuneiform tablets speak of field workers scooping up the pasty substance into

leather or reed baskets and shipping them by boat convoys in quantities sometimes exceeding 3 000 pounds. Yet even with this bounty at hand, there were times when supply could not keep up with demand; and in times of temporary shortages, tempers flared up. *"Concerning ½ ounce of bitumen"*, wrote one disgruntled merchant of 2500 BC to his supplier, *"for the fifth time I have written to you and you do not send it. Find me ¹/₂ ounce and send it!"*.

EARLY GLASSWARE, OBJECTS WITH ARTIFICIALLY GLAZED SURFACES ARE KNOWN TO HAVE BEEN MADE BEFORE 3000 BC.
THE MANUFACTURE OF GLASS VESSELS APPEARS TO HAVE ORIGINATED IN MESOPOTAMIA ABOUT 1600 BC, AND USED THE TECHNIQUE OF 'LOST WAX'
CASTING BEFORE GLASS BLOWING HAD BEEN INVENTED. BITUMEN MAY HAVE BEEN USED TO FUEL THE FURNACES.

GLASS JAR, BELONGING TO KING SARGON ABOUT 720 BC
FROM THE NORTH WEST PALACE AT NIMRUD.

Abu Ghir, the Arabic name of a town South of Hit, is one hint we have today of the past riches of this region. Ghir meaning a mixture of bitumen, sand and lime that was used by the Arabs well into the first half of this century in paving the roads of Baghdad and Basra.

Even after the destruction of Babylon, however, Hit's reputation as the oil capital of Babylonia remained strong. The town continued to provide vital fuel and petroleum cement to southern Iraq for many centuries thereafter. This

is supported, for example, by the testimony of the Roman writer Diodorus Siculus from the first century BC, who asserted, perhaps not without awe: *"although the sights to be seen in Babylonia are many and singular, not the least wonderful is the enormous amount of bitumen which the country produces; so*

great is the supply of this, that it not only suffices for their buildings, which are numerous and large, but the common people also, gathering at the place Hit, draw it out without any restriction and after drying it, burn it in place of wood. Countless as the multitude of men who draw it out, the amount remains undiminished, as if derived from some immense source."

Hit's oil and gas pits, which had dazzled long-forgotten emperors such as Tukulti-Ninurta, Nabopolassar and Nebucheddnezzar, also fascinated more familiar Greek and Roman names.

CONTEMPORARY MARBLE BUST OF THE
EMPEROR TRAJAN (AD 98 TO 117)
TRAJAN AND HIS AGGRESSIVE AMBITIONS WERE
RESPONSIBLE FOR ENDING THE NABATAEAN
EMPIRE, WHEN HE INCORPORATED IT INTO THE
PROVINCE OF ROMAN ARABIA IN AD 106.
HIS CAMPAIGN REACHED THE SHORES OF THE
PERSIAN GULF IN AD 116.

(PREVIOUS PAGES) THE FINDING OF MOSES
A BRONZE PLAQUE BY MASSIMILIANO
SOLDANI (1658 - 1740).
THE BIBLICAL STORY OF MOSES HAS
REMARKABLE SIMILARITIES TO THE POEM OF
SARGON, KING OF ASSYRIA RECORDED 2300
YEARS EARLIER.

"Sargon, the mighty King of Agade, am I.
My city is Azupirani, which lies on the
bank of the Euphrates.
My humble mother conceived me,
she brought me forth in secret.
She laid me in a basket made of reeds,
she smeared the door with bitumen
and committed me to the river which did
not submerge me.
The river carried me to Akki who made me
his gardener. The Goddess Ishtar fell in
love with me and for fifty-four years I
ruled the kingdom".

KING SARGON (2300 BC) RULED OVER MUCH
OF ANCIENT MESOPOTAMIA FOR MORE THAN
50 YEARS AND IS KNOWN TO HAVE BUILT A
ROYAL PALACE AT AKKAD. THE CITY OF AKKAD
HAS LONG SINCE BEEN COVERED BY THE
DESERT AND ITS LOCATION AND TREASURES
ARE YET TO BE DISCOVERED.

Alexander the Great, Trajan, Severus, and Julian are all Western emperors who had visited the town and watched its wells spew out the mysterious, magical oil that they - unlike the Near Eastern people - could not fully appreciate. Most of the oil from Hit and the other smaller sites went into building and maintaining what may be called the 'infrastructure' of the Sumero-Babylonian civilisation. This included the waterproofing of boats, the construction of massive monuments, the paving of roads, the laying of sewers, the reclamation of land and the construction of dykes and canals. In fact a shortage of crude oil mastic in these areas would have been nearly as devastating in its socio-economic consequences as a shortage of building cement would have been in our own times. Still, not all the oil was used in construction projects; a small portion found its way into fields which,

though unrelated to infrastructure, were nevertheless of crucial importance to society. Medicine and the fine arts are two good examples.

In the field of medicine, the ancient inhabitants of Iraq - like most ancient people - believed that sickness and disease were caused by malevolent spirits, or devils. Therefore, incantations and spells were believed to play just as important a role in the cure as any herbal prescriptions. These speak of the common practice of carving the face of the sick person on a small block of rock-hard bitumen and then, after reciting an exorcising spell, tossing the carving in the Euphrates or Tigris as a symbol of deliverance. Often, such exorcising spells would use the opening sentence: *"Like bitumen and pitch which comes out of the depth…."* or *"As bitumen holds a ship, so I hold you and will not let you go…"*

EKIM DOCTOR.
FOR THOUSANDS OF YEARS DOCTORS PRESCRIBED BITUMEN AS A CURE FOR MANY AILMENTS.

WATER PIPE REPAIRED WITH BITUMEN
FOUND AT TELL AL - UBAID,
DATING FROM 4000 BC.

Sorcerers augured by the appearance of crude oil as it floated into a dish of water: "*If when I pour naphtha on water it has the appearance of bitumen, that means fate, the sick man will die.*" Or "*If there are agglomerations like naphtha or sesame oil borne hither and thither on the surface of the water, disease will lay hold of the land.*" If a person was sick and all else failed, the door to their home was painted with bitumen to ward off the evil spirit. Or, if his affliction was dropsy, he was made to stand on a block of bitumen.

Not all healers, however, were witches and sorcerers. Some were physicians in the sense that they tried to diagnose symptoms rationally and prescribe appropriate cures. Treatments included warm bitumen as an ointment for swollen feet and hands; bitumen or naphtha as a wound healant; a mixture of bitumen and olive oil for soothing the eyes; and a beverage of bitumen, beer and various herbs prescribed as a sedative for the stomach:

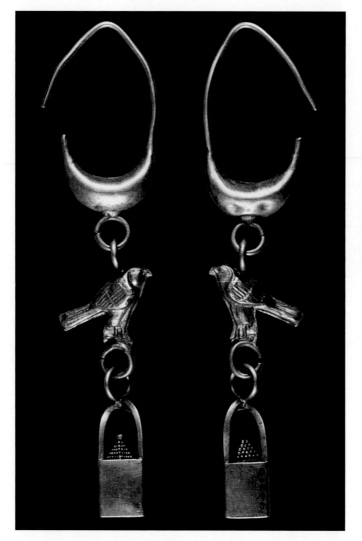

GOLD EARRINGS,
FROM THARROS
700 BC.
MODELLED IN THE
SHAPE OF A HAWK
CARRYING A BASKET.

GOLDEN WIG, FROM UR 600 BC.

"Having crushed the roots of (such and such) plants together with dried river bitumen .... and having massaged (the affected area) with oil, you shall put (the preparation) on as a poultice."

Crude oil mastic was also an important medium for artistic expression in ancient Iraq. In fact some of the oldest and most stunning artistic objects found

(ABOVE) "...TO WALK ON WATER WITH BITUMEN SANDALS".

below the sands of the Euphrates-Tigris plain - from religious and political artefacts to the most exquisitely made jewellery - testify not only to the sophistication of Sumerian and Babylonian artistic design and craftsmanship, but also reaffirm how pervasive the use of petroleum was in that ancient land.

(LEFT) EASTERN COOKING SCENE.
THE ART OF BLENDING MAGIC WITH SCIENCE WAS PRACTISED AS LONG AGO AS 2500 BC. ANCIENT DICTIONARIES WRITTEN ON STONE TABLETS TELL OF HOW SUCH PRACTICES PROTECTED PEOPLE FROM THE DEMONS WHO THEY BELIEVED WERE RESPONSIBLE FOR DISEASE AND FAMINE. SUCH BELIEFS INFLUENCED IMPORTANT DECISIONS OF STATE.

A DERVISH.

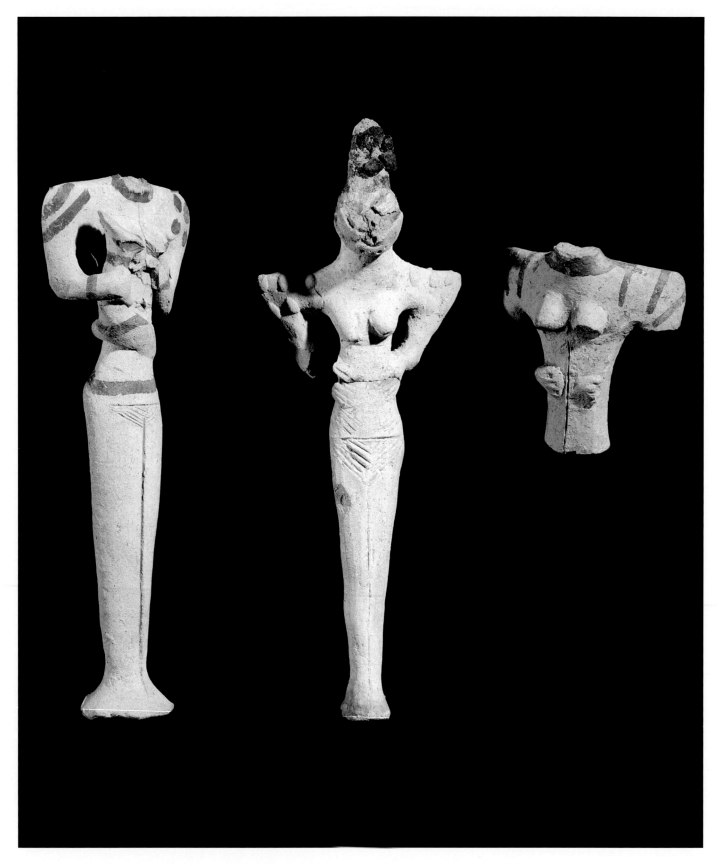

CLAY GODDESS FIGURINES WITH BITUMEN WIGS 5500 to 4000 BC.
MADE DURING THE UBAID PERIOD PROBABLY FOR MAGICAL OR RELIGIOUS PURPOSES.

DETAIL OF FRIEZE SHOWING BULLS, FROM TELL AL-UBAID 2500 BC.
THE WHITE STONE ANIMALS CONTRAST WITH THE BITUMEN BACKING.

Between 4000 and 3000 BC, for example, bitumen was moulded into wigs to adorn terracotta figurines and larger stone statues representing the fertility goddesses. It was also used as a base in which to inlay precious stones and small fragments of shell for decorating ornamental objects intended for household or ritual use, such as jewellery boxes, toys and musical instruments.

One of the most famous ornamental objects found in ancient Iraq was from a royal grave at the cemetery of Ur. It consists of a standing goat which may have served to support a piece of furniture used during prayer. The head and legs of the goat, as well as the thicket on which it is leaning, are carved out of wood and covered with gold foil, cemented with bitumen.

OIL OMEN TABLET 2000 TO 1700 BC.
THE TABLET WAS USED TO CALCULATE OMENS FROM PATTERNS CREATED WHEN OIL WAS DROPPED INTO WATER.
...Sorcerers augured by the appearance of crude oil as it floated into a dish of water:
"If when I pour naphtha on water it has the appearance of bitumen, that means fate, the sick man will die".

A FUNERAL CEREMONY FOR A KING.
THE PEOPLE OF ANCIENT MESOPOTAMIA HELD ELABORATE FUNERAL CEREMONIES.
IN ORDER TO SUSTAIN THEMSELVES IN THE AFTERLIFE ITEMS OF PERSONAL WEALTH
(INCLUDING BITUMEN) WERE BURIED WITH THEIR REMAINS.
MUCH LATER THE PHARAOHS OF EGYPT ADOPTED SIMILAR BELIEFS.

The back and flanks of the animal were also coated with bitumen in which shells had been embedded, and the horns and eyes of the animal, which symbolised fertility, were of lapis lazuli.

Another famous item from this cemetery is the so-called 'Standard of Ur', discovered in a plundered royal tomb dating to about 2500 BC.

The 'Standard's' two main sides are wooden panels about 18 inches wide which depict in vivid details, scenes of Sumerian life. One panel shows a king holding a festive banquet while his subjects line up in large numbers to bring him gifts of livestock and produce; the other panel shows a warfare scene with a mighty army decimating the enemy and taking them prisoner to show to the proud king.

ROYAL STANDARD OF UR 2600 TO 2350 BC.
A HOLLOW BOX WHOSE ORIGINAL FUNCTION IS NOT KNOWN, THIS WAS POSSIBLY PART OF A MUSICAL INSTRUMENT.

(PREVIOUS PAGES AND ABOVE) *"…the 'Standard of Ur,' dating to about 2500 BC.*
The 'Standard's' two main sides are wooden panels about 18 inches wide which
depict in vivid details scenes of Sumerian life".

ROYAL STANDARD OF UR.
DETAIL SHOWING FIGURES OF SHELL AND LAPIS LAZULI INSET ON A BASE OF BITUMEN.

73

WOMAN'S HEAD-DRESS FROM UR 2600 BC.
CORNELIAN AND LAPIS BEADS, WITH GOLD LEAF DROP PENDANTS.

JEWELLERY FROM UR 2600 BC.

The much later fall of Babylon into the hands of Cyrus the Great, King of Persia in 539 BC, marked the end of what may be called the first golden age of petroleum, the Sumero-Babylonian age. Demand for crude oil declined considerably in Iraq after that date partly because of the general decline of the Babylonian civilisation.

Between the years 400 BC and 300 BC the focal point of the oil trade shifted from the pits and smoke-holes of Babylonia to the briny depths of the Dead Sea, for by that time oil had become the hottest item of trade between the Egyptians and a lesser known band of nomadic who made their living catching 'bulls' in the great salty lake.

CHAPTER
TWO

———

BULLS
FROM THE
SEA

Bitumen carving, from Susa 600 bc.
Showing a woman spinning wool.

For the ancient Arabian people known as the Nabataeans, history began in 312 BC when an army of Greek mercenaries crossed the Syrian Desert into present-day Jordan and headed toward the Southern tip of the Dead Sea. When they reached their destination their commander - a general named Hieronymus of Cardia - was astounded by what he saw. Scores of Arab-speaking tribesmen were camping on the shore with pack-camels couched and reed rafts beached, waiting for what they called the 'thawr' - Arabic for bull - to appear in the middle of the sulphur-smelling waters. Every time a new 'bull' came into sight, a swarm of

axe-wielding seamen leapt into their rafts and began a frantic race toward the catch.

Stone tablet, Mesopotamia
Uruk period 3300 bc.
Its purpose is not fully understood
but this tablet may record a sales
transaction and depict the people
involved.

Amazingly, the catches were neither bulls nor sea-creatures of any kind. They were mounds of jellied crude oil -

or bitumen - that erupted like mini-icebergs from the depths of the murky sea and drifted aimlessly with the wind. It was obvious that the Arabs prized the oily exudate immensely; as the Greeks put it, they carried it off like 'plunder of war'. Nowhere is this scene more vividly depicted than in Hieronymus' own journal.

"They make ready large bundles of reeds and cast them into the sea. On these not more than three men take their places, two of whom row with oars, which are lashed on, and one carries a bow to repel any who sail against them from the other shore or who venture to interfere with them.

(previous pages) "…every time a new 'bull' came into sight, a swarm of axe-wielding seamen leapt into their rafts and began a frantic race toward the catch".

GOLDEN STATUETTES, FROM THE OXUS TREASURE, NORTHERN BACTRIA 500 BC.
MINIATURE BEARDED MEN WEARING HIGH STIFF CAPS, TUNICS AND EMBROIDERED COATS
WITH EMPTY SLEEVES, HOLDING BUNDLES OF SACRED RODS.

PETRA - THE
'ROSE RED CITY'
ANCIENT CAPITAL
OF THE
NABATAEANS -
LIES HIDDEN IN
THE MOUNTAINS
OF THE RIFT
VALLEY WHICH
RUNS BETWEEN
AQABA AND THE
DEAD SEA.

GOLD BIRD'S HEAD ORNAMENT, FROM THE OXUS TREASURE, NORTHERN BACTRIA 500 BC.
GOLD ORNAMENT IN THE FORM OF A STYLISED BIRD'S HEAD.

When they come near the floating bitumen they jump upon it with axes and, just as if it were soft stone, they cut pieces and load them on the raft, after which they sail back."

On the shore, crews of women and children sprinkled the hunks of tarry oil with sand, stuffed them into leather bags and loaded them onto camels for the long journey across the Negev and Sinai.

Their final destination: Alexandria, Egypt. Ironically, when Hieronymus witnessed this extraordinary harvest of the sea, he was under orders to expel the Arabs and secure the oil for his master, the Macedonian Greek King Antigonus I Monophthalmos - the 'One-Eyed'.
As it happened, Hieronymus' talent for keeping good notes of his observations far exceeded his skills as a military leader.

His army was heavily defeated, and he had to flee back to Syria for his life. Some 270 years later, his diary, long forgotten, fell into the hands of the famous Roman historian Diodorus Siculus who made good use of it in his description of early Nabataean life. Until that eventful day when the Greeks made their unwelcome appearance in their midst, almost nothing was known about these

Detail of the Queen's Lyre showing the golden head of a mystical bull.

mysterious oilmen of the Dead Sea, the Nabataeans. In recent decades, archaeologists have established beyond doubt that their home base was the present-day Hijaz region in North-Western Saudi Arabia, from which they eventually fanned out to build a kingdom that included large tracts of the Negev Desert, almost all of what is now Jordan, and at one time, even Damascus. Harvard Professor G.W. Bowersock, an expert on the Nabataeans and author of the book 'Roman Arabia', called their realm *"one of the greatest kingdoms of the ancient Near East."* The Nabataeans were known to the Greeks and Romans as a nation of wealthy traders - so wealthy in fact that they are the only people in history known to have imposed a punitive tax on whomever among them grew poorer instead of richer. To be sure, much of their fabulous wealth came from their tight grip on the caravan trade in spices and incense from Southern Arabia to Egypt - a lucrative trade for which they were the exclusive middlemen.

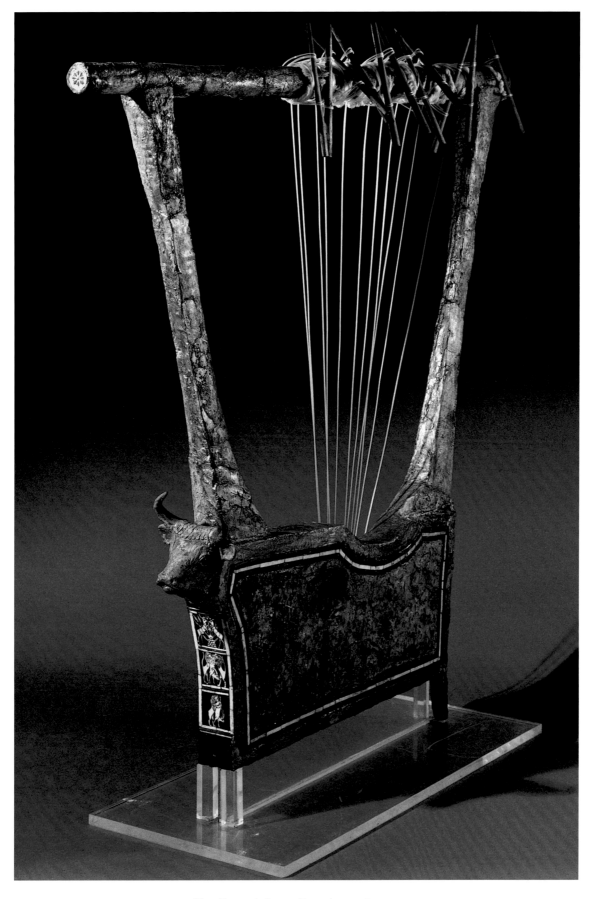

THE QUEEN'S LYRE, SUMERIAN 2600 BC.
DECORATED WITH LAPIS LAZULI, SHELL AND RED LIMESTONE MOSAIC, SET IN BITUMEN.
THE HAIR AND BEARD ARE OF LAPIS LAZULI, THE SIGNIFICANCE OF THE BEARD IS NOT KNOWN.

A strong case can be made that these shrewd and studious Arabs built their prosperity on two monopolies, not one, for they were also the sole exporters of Dead Sea oil to the richest and most populous country of its time, Egypt.

Petroleum, and in particular bitumen, had never played a significant role in the civilisation of ancient Egypt, at least not until a few centuries before the Christian era. Even when the country was at the pinnacle of its power, around 3000 BC, Egyptian writings hardly ever mentioned the substance or listed it as one of the imperial imports. This, despite the fact that commercial ties between Egypt and ancient Iraq - the largest oil producer, consumer and exporter in antiquity - were often very extensive. There were reasons, of course, why this was the case. Foremost among them was that the country itself lacked any significant source of oil - nothing, in any case, that compared in size with the oil seepages of Iraq for example.

'THE ENCAMPMENT OF THE AULAD-SAID'
- DAVID ROBERTS RA (1796 - 1894).
"...The Nabataeans' fabulous wealth came from their tight grip on the caravan trade in spices and incense from Southern Arabia to Egypt".

Although some oil seeps may have been known in the Sinai Peninsula, they were never substantial, and in any case may never have been exploited before the Greco-Roman period. Second, the Egyptians were never as preoccupied with waterproofing as the mudbrick masons or the reed shipwrights of the Euphrates. The flood waters of the Nile, always predictable and slow in their annual rise and recess, were never a threat to their giant pyramids. These, after all, were built not of crumbly mudbricks as was the case in Babylonia, but of massive cut stone weighing at times more than two tons a piece. Limestone provided the Egyptian builder with an excellent source of cement in the powder form, something that the ancient inhabitants of the Tigris-Euphrates valley never had.

(PREVIOUS PAGES) THE STREET BAZAAR.

"...The flood waters of the Nile, always predictable and slow in their annual rise and recess, were never a threat to their giant pyramids".
BITUMEN WAS THEREFORE NOT A VITAL BUILDING MATERIAL.

Finally, the ancient Egyptians did not need crude oil mastic to caulk their ships. They used instead pitch made from the wood they imported in large quantities from the mountains of Lebanon and from other parts of Africa. The situation however, had changed dramatically by the time the venturing Greeks first launched their expeditions against the Nabataeans. For by then, Egypt's imports of bitumen from the Dead Sea had grown significantly - in fact, significantly enough to become the focus of war and historical records.

The substance was fast becoming the main ingredient in the Egyptians' most important religious ritual - mummification. Our chief informant on this matter was Diodorus Siculus who lived around 100 BC.

MUMMY CASE OF
DJED-BASTET-IUF-ANKH 100 BC.
COFFIN LID DECORATION SHOWING THE
MUMMIFICATION RITUAL.

EMBALMERS POURING
WATER FROM JUGS.
THE BODY, BLACK FROM
OILS AND RESINS IS
PURIFIED WITH STREAMS
OF WATER.

HEAD EMBALMER WEARS
AN ANUBIS MASK.
THE BODY LYING ON A
COUCH IS COVERED IN
DRIED CRYSTALS OF
NATRON.

CANOPIC JARS.
THE MUMMY BANDAGED
AND WEARING A MASK IS
ATTENDED BY ANUBIS.

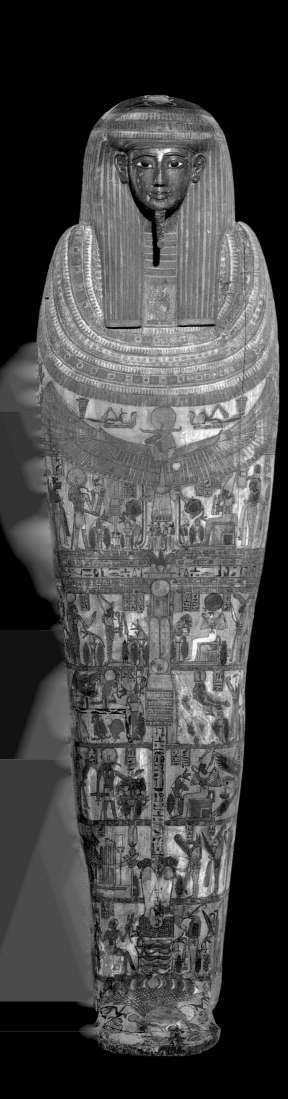

THE EGYPTIANS BELIEVED THAT THE PERSON'S SOUL LEFT THE BODY AT DEATH. AFTER THE BURIAL, THE SOUL WAS REUNITED WITH THE BODY IN THE AFTERLIFE. FOR ALL THIS TO HAPPEN, THE BODY HAD TO BE WELL PRESERVED. MUMMIFICATION REACHED ITS PEAK AROUND 1000 BC, BUT ROMANS LIVING IN EGYPT WERE STILL BEING MUMMIFIED IN AD 300. THE ADOPTION OF MUSLIM AND CHRISTIAN BELIEFS ENDED THE LONG TRADITION OF MUMMIFICATION AND THE ANCIENT EGYPTIANS' DEMAND FOR BITUMEN.

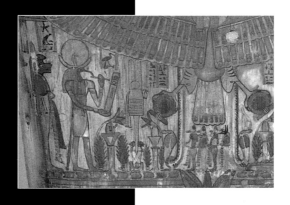

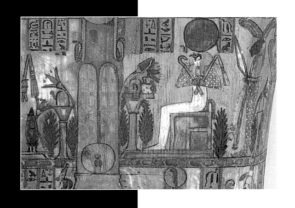

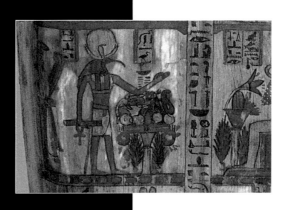

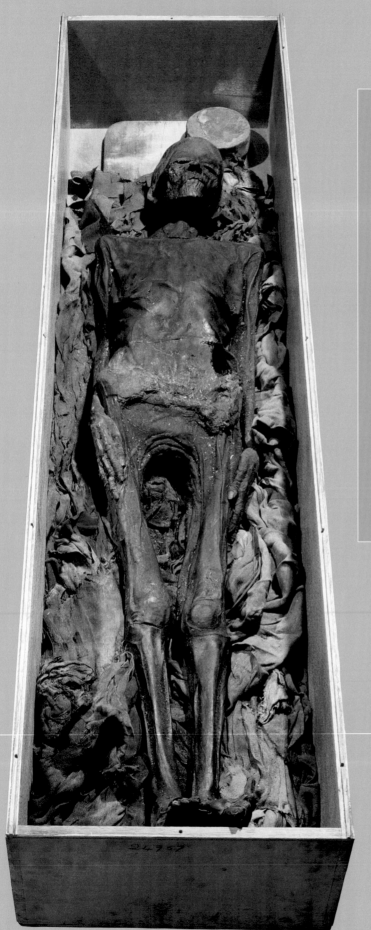

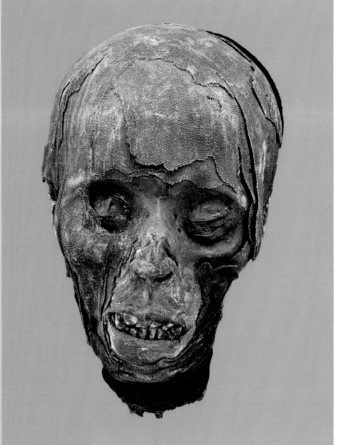

(ABOVE) MUMMIFIED HEAD 600 BC.

(LEFT) MUMMIFIED WOMAN 600 BC.
THE WELL PRESERVED MUMMY OF A WOMAN
FROM AROUND THE TIME HERODOTUS VISITED
EGYPT. THE APPEARANCE OF THE MUMMIFIED
BODY HAS A REMARKABLE RESEMBLANCE TO
NATURAL BITUMEN. THE WORD 'Mumija'
DENOTES WAX IN PERSIAN AND BITUMEN IN
ARABIC.

RECENT SCIENTIFIC ANALYSIS OF MUMMIFIED REMAINS HAS REVEALED THAT THE ANCIENT
EGYPTIANS ACQUIRED BITUMEN FROM AS FAR AWAY AS HIT IN MODERN-DAY NORTHERN IRAQ,
EVIDENCE OF THE EARLY TRADE THAT EXISTED FOR THIS PRECIOUS NATURAL RESOURCE.

Referring to the Nabataeans as barbarians - because they were not Greek subjects - he explained that *"the barbarians who enjoy this source of income take the asphalt to Egypt and sell it for the embalming of the dead; for unless this is mixed with the other aromatic ingredients, the preservation of the bodies cannot be permanent."* Embalming of the dead had been an Egyptian tradition for thousands of years driven by the ancient Pharaonic belief that, without preservation of the body, there could be no assurance of an afterlife for the soul.

However, sometime around 500 BC, the Egyptians began to experience a shortage of the aromatic resins, gathered from shrubs and trees, that they relied upon for the operation. Eventually, Egyptian priests found that they could do just as well by reducing the quantities of these aromatics and mixing them with molten bitumen - or *mrhe.Hr*. This word is translated from an ancient Egyptian papyrus text on embalming written before 1500 BC as a supplication to Anubis, the jackal-headed funerary deity and reputed inventor of mummification: *"Anubis..... fills the interior of the skull with mrhe.Hr, incense, myrrh, cedar oil, and calves' fat"*.

Gold horn filled with bitumen, from Tell al-Ubaid 2500 bc.
Part of a golden bull from the temple of Ninhursag, the great mother goddess.
The facade of the temple was decorated with an extensive amount of gold,
this is the only piece to have survived.

The word *mrhe.Hr* is written in hieroglyphics with the signs owl, mouth, twisted flax, face and pot. Although the word is suspiciously close to the Arabic name for bitumen, *humar*, which the Nabataeans may have used, there is in fact no proof that the two words are connected. Egypt had no significant bitumen or oil deposits of its own and had to import the substance from the Nabataeans. Egypt's population around 300 BC was close to seven million, demand for bitumen was very large, and the Nabataeans knew that in the oil of the Dead Sea they had their hands on a fortune of immense proportions. Then as now, however, oil and international politics proved to be inseparable. Between 323 and 285 BC the Nabataeans suddenly found themselves at the centre of a bitter struggle between two superpowers each headed by a former general in the army of Alexander the Great: Ptolemy I Soter and Antigonus I Monophthalmos.

After the death of Alexander, each of the generals aspired to eliminate the other and carve out for himself an empire that would include all of the Near East. Ptolemy I founded a dynasty in Egypt, whereas Antigonus I retained part of present-day Turkey and all of Syria and Lebanon.

TOMB OF EGYPTIAN KING SETI, DEPICTING THE SEATED OSIRIS,
PERSONIFICATION OF THE PHAROAHS, WITH HIS SON THE
FALCON-HEADED HORUS.
THE EGYPTIAN MYTH BEHIND THESE CHARACTERS TELLS HOW OSIRIS WAS
KILLED BY HIS BROTHER SET, ONLY TO BE RESURRECTED BY THE MOTHER
GOD ISIS AS THEIR SON HORUS.
OSIRIS ALWAYS APPEARS AS A BLACK GOD AND, PERHAPS, WAS
ILLUSTRATED AS SUCH BECAUSE THE PHAROAHS ATTEMPTED TO
REPRODUCE HIS RESURRECTION IN THEIR OWN MUMMIFICATION
INVOLVING BLACK BITUMEN.

TRIPOD OF BITUMEN MASTIC FROM SUSA 2000 BC.

VESSEL MADE FROM BITUMEN WITH LONG HANDLE,
PROBABLY USED AS A DRINKING CUP, FROM SUSA 2000 BC.

COIN SHOWING PTOLEMY I SOTER.
A GENERAL IN THE ARMY OF ALEXANDER THE GREAT,
WHO FOUNDED A DYNASTY IN EGYPT AFTER HIS DEATH.

In the year 312 BC, Antigonus made his move against Ptolemy: he dispatched a trustworthy officer, Athenaeus, at the head of an army of 4 600 men with the dual mission of subduing the 'barbarians', as the Greeks referred to the Nabataeans, and of imposing an economic blockade against Egypt's Eastern flank. Antigonus had also heard of the bitumen exports across the desert and may have calculated that a shortage of the substance in Egypt would almost certainly stir up the Egyptian priesthood against his rival. But Antigonus' scheme failed. Informers had told one of his officers that it was a custom of the nomadic men of the Dead Sea area to gather once a year for a national festival, during which they left all their possessions, old people, women and children for safekeeping at a certain place referred to as 'The Rock'.

This rock, described as being exceptionally strong, high, and unprotected by a wall - sounds very much like the hill Umm al-Biyarah, within the rose-red city of Petra, whose name means 'rock' in Greek, and which was later to become the Nabataean capital. Athenaeus timed his raid to coincide with the festival. Reaching the rock at nightfall, he surprised the Arabs there and killed or imprisoned many of them.

After looting the encampment he made off in the darkness with 700 camels, and a large quantity of booty that included much frankincense and myrrh and about 500 talents of silver. To us, such riches are evidence that the Nabataeans had by that time accumulated great wealth from their trade, but for the forces of Athenaeus they spelt doom. Weary in the torrid heat, heavy-laden and short of water, they made the mistake of setting up camp too soon. Having learnt of the Greek assault the Nabataeans meanwhile, lost no time in gathering their forces and sending them in pursuit of the intruders. Locating the Greek camp, they fell upon it with vengeance and at the end of the day all of Athenaeus' infantry and most of his cavalry were destroyed. When they returned to their rock, the Nabataeans sent an angry letter to Athenaeus accusing him of aggression, and demanding assurances that the Greeks would not harm them again. Antigonus, hoping to buy time, denied all responsibility for the incident and assured them that his army had acted against orders. Shortly thereafter, Antigonus summoned his own son Demetrius - known to the Greeks as Poliorcetes, 'the Besieger' or 'City Sacker' - to march on the rock at the head of a formidable force of more than 8 000 troops. His movements, however, were detected very early on, and news of the imminent invasion reached Petra by fire signals.

When Demetrius reached Petra, he found it heavily guarded and his first assault was easily repelled. The next day, as he was preparing to storm the city a second time, the Nabataean elders sent him a message that remains to this day one of the most beautiful expressions of the Bedouin spirit: *"King Demetrius, with what desire or under what compulsion do you war against us who live in the desert, and in a land that has neither water nor grain nor wine nor any other thing whatever of those that pertain to the necessities of life among you.*

(PREVIOUS PAGES) THE TOMBS AT PETRA - DAVID ROBERTS RA (1796 - 1894).

CUPS MADE FROM OSTRICH EGGS, FROM UR 2600 BC.
THESE HAVE BEEN INLAID WITH STONE, SHELL AND MOTHER OF PEARL ON A BITUMEN BASE.

For we, since we are in no way willing to be slaves, have all taken refuge in a land that lacks all the things that are valued among other peoples, and have chosen to live a life in the desert and one altogether like that of wild beasts, harming you not at all.
We therefore beg both you and your father to do us no injury but, after receiving gifts from us, to withdraw your army and henceforth regard the Nabataeans as your friends. For neither can you, if you wish, remain here many days since you lack water and all the other necessary supplies, nor can you force us to live a different life."

Demetrius finally agreed to withdraw, on the condition that he be given hostages and gifts. Instead of retreating toward Syria, however, Demetrius - to the Arabs' dismay - marched on to the Dead Sea and declared himself 'Lord of all the oil fisheries'. Leaving behind his army, he then hastened back to his father to report the news.

LADY LAYARD'S JEWELLERY, ORIGINATES 2300 BC.
A UNIQUE ITEM OF JEWELLERY COMMISSIONED BY VICTORIAN ARCHAEOLOGIST AND 'ASSYRIAN ADVENTURER' SIR AUSTIN HENRY LAYARD, AND PRESENTED AS A WEDDING GIFT TO HIS WIFE ENID. MADE FROM CYLINDER SEALS AND CORNELIAN RECOVERED FROM HIS EXCAVATIONS IN MESOPOTAMIA. LADY LAYARD LATER WROTE IN HER DIARY THAT WHEN SHE DINED WITH QUEEN VICTORIA IN 1873 "IT WAS MUCH ADMIRED".

Antigonus, apparently impressed by what his son had achieved, immediately sent another battalion to the Dead Sea, headed this time by Hieronymus, with specific orders to "*prepare boats, collect all the bitumen, and bring it together in a certain place.*" However, when Hieronymus attempted to harvest the oil with his boats, his forces were attacked by no less than 6 000 Arabs - some of them on rafts - and were annihilated in a shower of arrows. The three campaigns of

Atheneus, Demetrius and Hieronymus in 312 BC marked the entry of the Nabataeans into recorded history. Clearly, by that time they were already rich and powerful. In the first half of the third century, during the reign of Ptolemy II Philadelphus (285-247 BC), the Nabataean territory expanded further

to include the area of the Hawran in present-day Syria, and a larger tract of the Negev desert across the Wadi Araba. Very little is known about the history of the Nabataeans from 350 BC to about 100 BC, except for the name of al-Harith (or Aretas) I, the first individually recorded Nabataean ruler, in about 168 BC, and the fact that, by around 200 BC, Petra was full of foreign dignitaries, including a Roman ambassador.

EARTHENWARE WATER JAR LINED WITH BITUMEN.
ORIGINALLY USED FOR CARRYING WATER, IN ANTIQUITY IT HAS BEEN ADAPTED INTO AN OSSUARY
(OR BONE URN) CONTAINING HUMAN BONES.

The Temple of Bacchus.

Silver gilt vase, Sassanian ad 600.
With scenes showing a grape harvest.

PRE-ISLAMIC ARABIANS REFERRED TO BITUMEN LINED EARTHENWARE AS 'MUZAFFAT' (FROM THE WORD *zift* MEANING PITCH OR BITUMEN). THE PROPHET MUHAMMAD SAW IN THE MUZZAFAT A SYMBOL OF ONE OF THE GREATEST SINS OF ISLAM *"Avoid the muzzafat; that bitumened cup"* HE REPORTEDLY DECLARED TO HIS FOLLOWERS.

We know, however, from the Jewish historian Josephus, that in 88 and 87 BC, the Greek Seleucid King Antiochus XII launched two separate campaigns against the Nabataean King Ubaydah I, in a determined effort to capture the Nabataeans' oil industry. Both invasions were crushed. The Nabataeans continued their petroleum exports to Egypt well into the first century BC, and their wealth, already considerable, continued to grow. The bitumen was carried along the Wadi Araba to Petra and then to the coastal city of Gaza. From Gaza, it was either loaded aboard ships bound for Alexandria, or taken along the Mediterranean coastline in fresh caravans to Egypt. As the first century BC progressed, both the Nabataeans and their captive market, the Egyptians, faced a serious threat from the Romans in Syria. In 62 BC, a Roman officer from Syria named Scaurus led an expedition against the Nabataeans

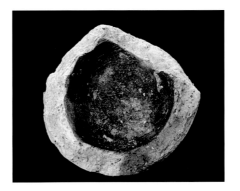

ASPHALT LINED JARS, FROM UR 1900 BC.

and withdrew only after being bought off with a large amount of silver. Seven years later another Roman commander, called Gabinius, invaded the kingdom and demanded another ransom in silver. The more silver the Nabataeans were

willing to part with, the bolder the Roman demands grew, until the Roman Triumvir Marc Antony, whose ambition was to expand and rule Rome's eastern dominions, finally annexed the Nabataean kingdom outright.

The Babylonians used cylinder seals to sign or stamp a document.
The seals were often made of semi-precious stone which was delicately carved.
Damp clay was used as the 'sealing' material and the seal was rolled over it
to form an impression.

Before long, however, Antony fell in love with that extraordinary Greek-Egyptian Queen, Cleopatra VII, who persuaded him to give her the Nabataean oil fisheries as a gift. To maximise her income from the operation while still ensuring that the oil would continue to flow into her kingdom, the astute queen contrived what may be considered as the first recorded lease-back scheme in the history of the oil industry. She leased the Dead Sea oil works back to the Nabataean King Malik I in 36 BC. Without any expenditure of money, labour or military force, Cleopatra thus assured herself a substantial revenue - which she intended to use to help build a naval fleet strong enough to defeat Antony's chief Roman rival, Octavian.

(previous pages) A Kanja on the Nile - A. Prisse d'Avennes (1807 - 1879). *"…the ancient Egyptians did not need crude oil mastic to caulk their ships. They used instead pitch made from the wood they imported from the mountains of Lebanon"*.

BUST OF CLEOPATRA VII, QUEEN OF EGYPT (69 to 30 BC).
"...the Queen contrived what may be considered as the first recorded
lease-back scheme in the history of the oil industry".

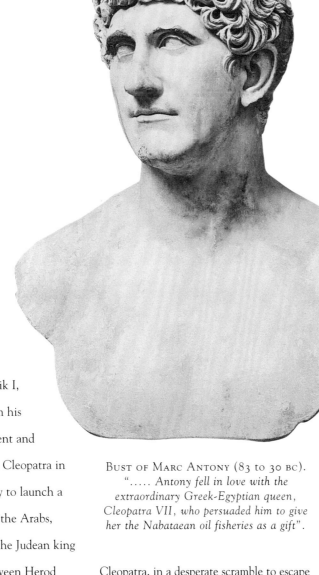

BUST OF MARC ANTONY (83 to 30 BC).
"..... Antony fell in love with the extraordinary Greek-Egyptian queen, Cleopatra VII, who persuaded him to give her the Nabataean oil fisheries as a gift".

The financial toll on the Arabs, however, proved unbearable: in 32 BC, Malik I, under intense pressure from his elders, balked at the payment and defied the Egyptian queen. Cleopatra in response, called on Antony to launch a punitive campaign against the Arabs, led by one of his protégés the Judean king Herod. The hostilities between Herod and the Arabs proved disastrous in the long run for the queen: the Arabs triumphed over Herod at the battle of Qanawat in present-day Syria, and shortly thereafter, in 31 BC, Antony himself was defeated by Octavian's forces in the naval battle of Actium, off the coast of Greece.

Cleopatra, in a desperate scramble to escape to India with Antony, had some of her ships dragged overland from the Nile to the Red Sea. But there, on the coast, her heavy-handed policy against the Arabs came back to haunt her: by an amazing coincidence of history, a garrison of Nabataeans was apparently stationed near the site where the ships were to be launched. No sooner were Cleopatra's galleys afloat than the Nabataeans - forever wary of an Egyptian naval presence on the Red Sea which might threaten their overland trade route monopoly - attacked and set them ablaze. Antony and Cleopatra were forced to flee back to Alexandria and the couple, realising that they were hopelessly trapped in Egypt, committed suicide. In a true sense then, the politics of oil in the ancient Near East sealed the fate of Antony and Cleopatra.

After the death of Cleopatra, Egypt became a colony of the new Roman empire created by Octavian, who took the title 'Augustus', and the Pharaonic custom of mummification ended.

Bronze of a young woman (200 to 100 bc).
Perhaps a priestess or goddess at the time of Cleopatra, she wears a hellenistic sleeveless tunic with a high girdle
and mantle. her jewellery consists of a torc and a bracelet on each wrist.
The defeat of Cleopatra and Antony at Actium in 31 bc may be taken as the political end of the hellenistic age.
Although its cultural influence was to last long into the period of the Roman Empire.

CLEOPATRA'S NEEDLE AT ALEXANDRIA - DAVID ROBERTS RA (1796 - 1864).
AN EXAMPLE OF THE MANY GIANT COMMEMORATIVE OBELISKS THAT WERE ERECTED BY THE ANCIENT EGYPTIANS.
LIKE SOME OF THE APPLICATIONS OF NATURAL BITUMEN, THE METHODS USED TO ERECT THESE HUGE STONES, WERE LOST OVER MANY CENTURIES.

As a result, the bitumen fisheries of the Dead Sea lost their economic importance. The Nabataeans retained some degree of independence from Rome until the year AD 106, when they were incorporated by the emperor Trajan into the newly formed province of Roman Arabia, with its capital at Bostra in Southern Syria.

Although the Romans did not practice embalming, they continued the Near Eastern custom - originally from Babylonia - of ingesting crude oil and bitumen for medicinal purposes, a custom that led more than a thousand years later, to one of the most bizarre trades in the history of mankind - the 'mumiya' trade.

"..... Egypt's imports of bitumen from the Dead Sea had grown significantly enough to become the focus of war and historical records .... the politics of oil in the ancient Near East sealed the fate of Antony and Cleopatra".

Ancient bitumen 'bulls' can still be seen on the shores of the Dead Sea.

Sometime in the 12th century, Medieval Europe learned that the ancient Egyptians had used a bitumen mixture called *mumiya* to embalm their dead, and that by melting this substance down one could obtain an oil that Ibn al-Baytar and other Arab physicians claimed was of great medicinal value. Beginning in that century, and throughout the Middle Ages, thousands of Egyptian mummies were exported from Alexandria to Europe by way of Marseilles - initially for recovery of the bitumen but later, as the effective ingredient was forgotten, to be simply ground to powder altogether. This powder, called 'mummy,' was a standard apothecary ingredient, and much in demand.

MUMIYA JAR.
THE GROUND UP REMAINS OF
EGYPTIAN MUMMIES KNOWN AS
'MUMMY POWDER' WERE A COMMON
APOTHECARY INGREDIENT IN
EUROPE FOR CENTURIES, THEIR
BITUMEN CONTENT BEING MUCH
VALUED.

MUMMY OF ARTEMIDORUS.
'AN EGYPTIAN OF GREEK BIRTH'. MODELLED IN RED STUCCO AND
COVERED WITH GOLD LEAF. MANY GREEKS AND ROMANS WHO
SETTLED IN EGYPT ADOPTED THE LOCAL PRACTICE OF MUMMIFICATION
AND THE LIFELIKE PORTRAIT IS CLEARLY THAT OF A GREEK RATHER
THAN AN EGYPTIAN.
OTHER LOCAL PRACTICES SUCH AS THE USE OF BITUMEN WERE ADOPTED
BY EARLY GREEK AND ROMAN VISITORS TO THE ANCIENT MIDDLE EAST.

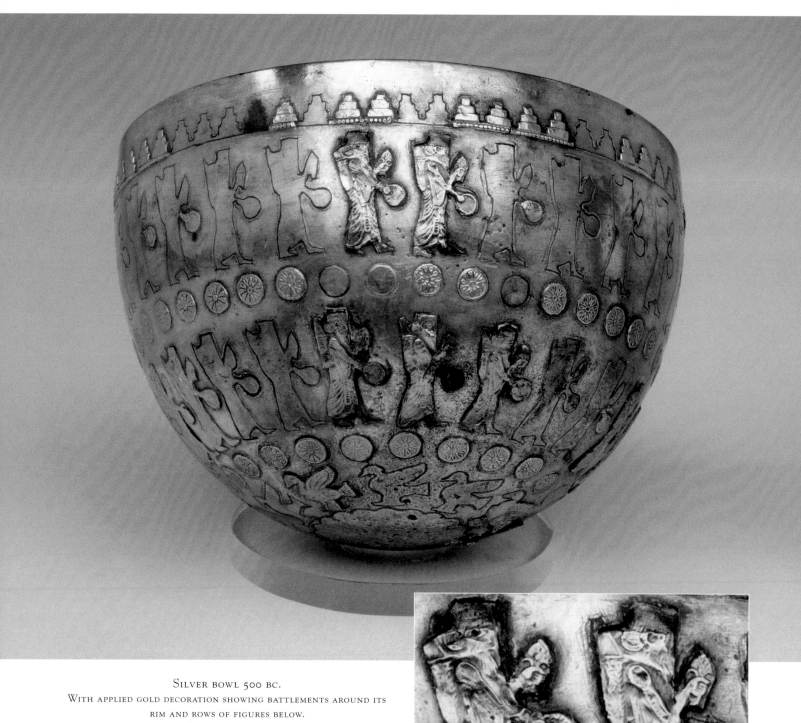

SILVER BOWL 500 BC.
WITH APPLIED GOLD DECORATION SHOWING BATTLEMENTS AROUND ITS
RIM AND ROWS OF FIGURES BELOW.

(PREVIOUS PAGES) PYRAMID TOMB OF LIPSIUS - PIERRE VAN ALVEN (1828-1908).

BITUMEN FOOT, AKKADIAN 2250 BC.
THE REMAINS OF A LIFE-SIZE BITUMEN STATUE.

When the supply of 'genuine' mummies and mummy parts ran low, unscrupulous gangs in Alexandria resorted to deceit, shipping instead the corpses of slaves and criminals which had been disembowelled, hastily aged in the sun and then stuffed or swabbed with a little bitumen, without proper cleansing. Not surprisingly, the practice had disastrous consequences in Western Europe, especially in France and Germany, where it contributed to the spread of the 'plague'. In 1564, a French physician, Guy de la Fontaine, travelled to Egypt and exposed the 'mummy' fraud to the French King. Unfortunately, his warnings were not immediately heeded, and the export trade in mummies did not come to a complete halt until some 50 years later. Indeed, domestically produced false 'mummy' was still traded in Europe around the year 1800 - a macabre offshoot of a Near Eastern petroleum industry of more than 2 000 years earlier.

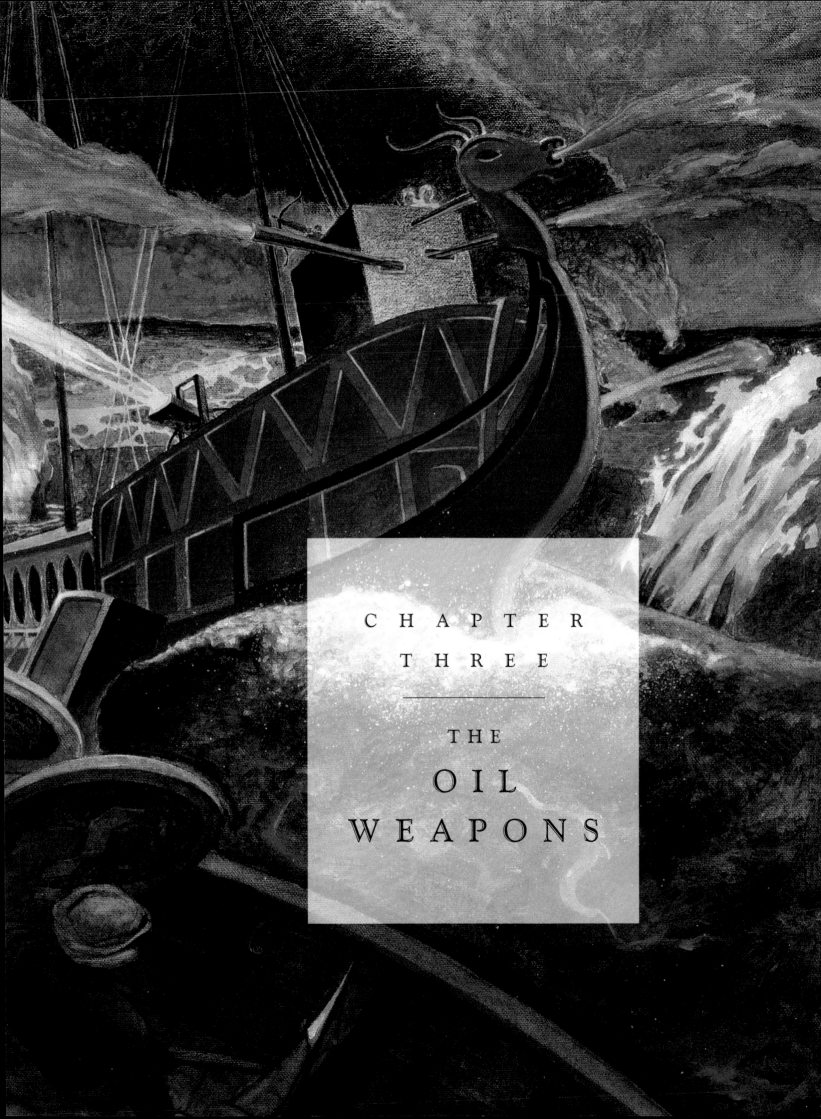

CHAPTER
THREE

———

THE
OIL
WEAPONS

PLANNING THE ATTACK - DAVID ROBERTS RA (1796-1864).

Sometime around the year 1400, a Persian scholar named Abu Tahir al-Fayruzahadi (AD 1329-1414) came to Mecca the holiest city of Islam - to perform the Hajj or pilgrimage and to write a lexicon of the Arabic language. Seven years later, he finished his monumental work entitled al-Kamus al-Muhit, or 'The All Encompassing Dictionary'. To this day, it remains one of the best references ever written on the Arabic language and culture, containing a remarkable section that deals with oil.

Not only does Abu Tahir dwell on the origin and nuances of the word 'naft' - then the Arabic word for naphtha which today means 'petroleum', but he also indicates that it was in common everyday use. A similar word 'Naffata' has three meanings: a naphtha well or fountain, a naphtha lamp used for lighting, or the brass instrument used to throw naphtha. In fact, we can legitimately infer from his writings all the elements of a thriving oil industry. Labourers evidently worked the oil wells, or naffata,

to obtain the oil, cameleers and merchants carried it and sold it in the cities as lighting fuel;

CEREMONIAL AXE HEAD FROM KERMAN, IRAN 2000 BC.

(PREVIOUS PAGES) "...galleys equipped with the 'fire-spouting devices' of the Byzantines."

WARRIOR OF THE NEGEV -
DAVID ROBERTS RA (1796-1864).

craftsmen built lamps and weapons of
brass or bronze to use in war; and
pharmacists made an assortment of
remedies from it.

But the most intriguing statement of Abu
Tahir is that 'water white naphtha' was
the best kind, implying that there was
another grade of inferior quality. Could
'water white naphtha' be the 15th-century
Arabic term for kerosene or another light
petroleum fraction, as opposed to the
darker crude oil? If so, then the Muslims
of that era must have practised some
form of crude oil refining and there must
have been refiners and other associated
technicians in this line of work.

(OPPOSITE) GOLD DAGGER WITH ELABORATE
SHEATH, FROM UR 2600 BC.

125

Many other Arab scholars before Abu Tahir had written in more detail on the subject, including physicians, historians, travellers, philosophers, military experts, alchemists and even poets. The Muslim Oil Age in fact began more than 700 years before Abu Tahir's book with a tale of early espionage. Sometime between the years AD 670 and AD 680, a discontented Syrian subject from Damascus sought refuge in Byzantium. Hardly anything is know about the identity of this man except that his name was Kallinikos, meaning 'handsome winner'. Kallinikos brought with him information that the Byzantine navy, besieged by the Muslims, welcomed very warmly. He was, as we would say today, a petroleum consultant and what he taught the Byzantines was no less than a secret formula for making something akin to modern-day napalm: a petroleum mixture that would burn anywhere even in water. The Emperor Constantine IV (AD 668-685) saw this as a chance to eliminate the Muslim threat to Constantinople. He ordered his high command to work with the defector in strictest secrecy in the construction of naphtha weaponry. In the seventh year of the siege, in AD 680, the fire of Kallinikos - later erroneously called 'Greek Fire' - was used in naval combat in what became known as the Battle of Kyzikos.

BATTLE SCENE FROM THE PALACE OF NINEVEH (NOW MOSUL IN NORTHERN IRAQ) 700 BC.

THE GARRISON RETURNS - S. BROUGH (1827 - 1878).

For the Muslim navy, the consequences were disastrous, Theophanes wrote. The entire flotilla, manned mostly by Syrians and Egyptians, was burned at sea, and Theophanes put the losses at 30 000 men.

The siege was broken, and the Muslims signed a 30-year truce. No doubt, the awesome naphtha cannons of the Byzantine navy - referred to as *mikroi siphones* by the Greeks and as *naffata* by the Arabs - were in widespread use in both navies in the ninth century. A crude picture of the Greek weapon is illustrated in a Byzantine manuscript, showing a ship from the fleet of the Emperor Michael I (AD 820-829). On the Arab side, the ninth century historian al-Maqdisi, stated that by AD 850, not only military ships, but also trading vessels in the Indian Ocean carried on board naphtha cannons to protect them against pirates in the open sea.

(OPPOSITE) STONE RELIEF, PALACE AT NINEVEH (NORTHERN IRAQ) 700 BC.
SHOWING SIEGE 'ENGINES' MOUNTING AN ATTACK ON A HEAVILY DEFENDED WALL. INSIDE SOLDIERS LADLE WATER TO PREVENT
BEING SET ALIGHT BY BURNING NAPHTHA TORCHES WHICH ARE THROWN DOWN FROM THE BATTLEMENTS ABOVE.

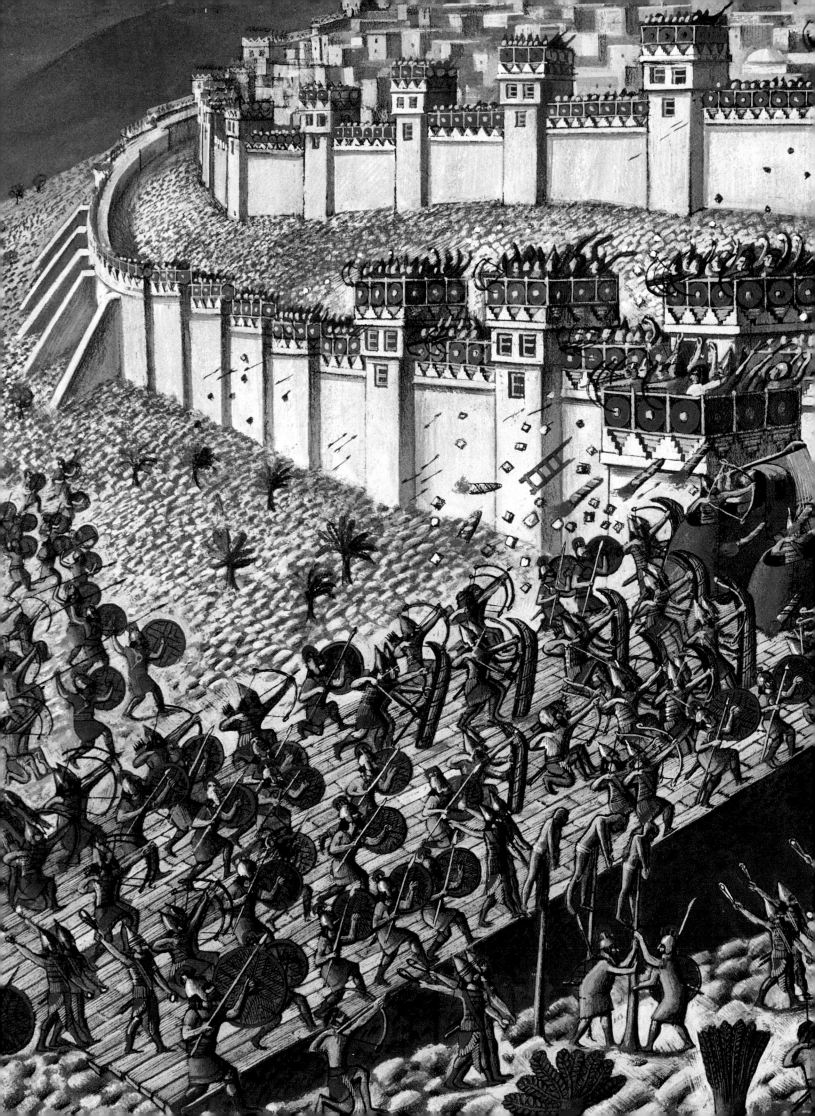

(LEFT) CEREMONIAL HELMET
AD 1400. INLAID WITH SILVER
PANELS OF PERSIAN VERSE.
(BELOW) ARMOURED GAUNTLET
AD 1500. DAMASCENED WITH
SILVER KORANIC INSCRIPTIONS.

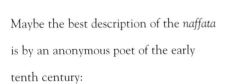

Maybe the best description of the *naffata*
is by an anonymous poet of the early
tenth century:

*" A yellow brass piece, with yellow viscous
spittle in its mouth.*

*When the spittle is thrown, it dances with the
wind and rivals it in quickness.*

From the bowels of the piece it emerges,

when released, disguised behind a dark cloak

that shields it like a protective fortress.

The piece and the head of the flying smoke,

each has a tail.

And when the fire is thrust forward, the

piece rushes backward as if suddenly stabbed

in the belly.

(PREVIOUS PAGES) RECONSTRUCTION OF THE SIEGE OF LACHISH, ASSYRIA.

THE SULTAN SELIM OF TURKEY.

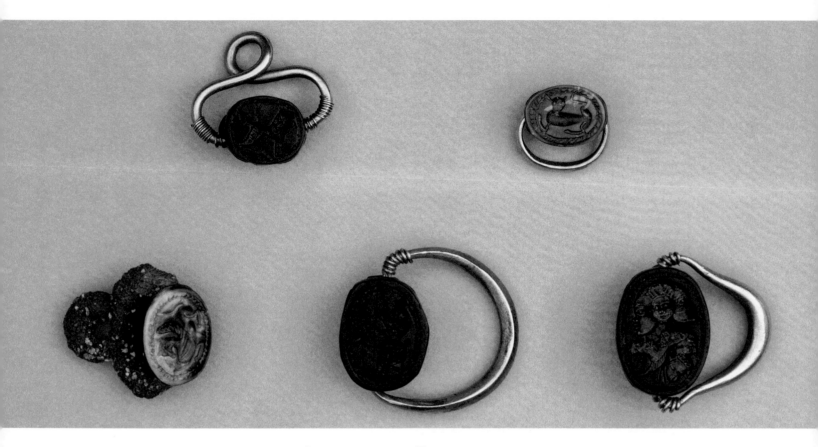

Scarab rings, from Tharros 700 bc.
Green jasper set in gold mount and swivel.

Waves of lightening between two nights [one night is the interior of the piece, the second is the black smoke], it sends from its belly to its mouth, toothless like the toothless mouth of a venomous snake.

And in the fray of battle it dives naked, and were you to ask her: are you afraid ?

It would know neither fear nor would she wish for safety."

There is no mention of oil, sea-fire or Kallinikos in any Muslim account of the battle of Kyzikos. The mere fact however, that Kallinikos fled Syria makes it more than likely that the secrets he took with him were already known to the Arabs, though as yet unadapted to their navy. By one account, however, when news of the battle at Kyzikos had reached the caliph Muawiyah in Damascus, he promptly sent word to his dockyards in Alexandria, home of some of the best shipbuilders in the Eastern Mediterranean, to equip his galleys with the 'fire-spouting devices' of the Byzantines. The extent of the use of petroleum-based weapons in the burnings of Medina and Mecca demonstrates that the Muslim forces had access to oil, were equipped to use it, transport it and deploy it anywhere in the empire. Whilst battle raged at Mecca in the summer of AD 683, a physician named Masarjawah, residing in the city of Basra in Iraq, was busy translating the first medical text ever put into Arabic for use as a manual with which to train the medics of the Muslim army. What resulted was the 'Book of the Powers of Remedies' or 'Kitab Qiwa al-Aqaqir', a collection of herbal recipes taken from a text originally written in Greek by an Egyptian priest. It was in this book that the term 'water white naphtha' was first used in Islamic medicine. How the Muslims obtained the oil is another story.

ARTIFICIAL FLOWERS, FROM TELL AL-UBAID 2500 BC.
FLOWERS OF POTTERY WITH STONE PETALS HELD IN PLACE BY BITUMEN.

CONSTANTINOPLE
HARBOUR.

136

"…Alexandria, home of some of the best shipbuilders in the Eastern Mediterranean".
THE PORT OF ALEXANDRIA - SITUATED AT THE MOUTH OF THE NILE, HAS PLAYED A VITAL ROLE IN TRADE
TO AND FROM THE MIDDLE EAST FOR CENTURIES.

SILVER SWORD HILT, FROM LURISTAN, WESTERN IRAN 800 BC.

When the Muslim armies first arrived in Iraq and Persia around AD 640, they found hundreds of open oil pits. Arab records from the time show that the province of Fars, in Persia, paid an annual tribute of 90 metric tons of oil to light the palace of the Caliph in Baghdad. Clearly the demand for oil was high.

Of these oil pits, the largest and most famous in Medieval times were at Jabal Barama, East of the Tigris in the North of Iraq, and the well of Dir al-Qayyara near Mosul. The caliph leased this latter well to private entrepreneurs and derived thousands of dirhams of annual revenue from them. So vast and strategically important was this pit of Dir al-Qayyara that at one time it had to be guarded day and night. It provided not only crude oil but most of the bitumen used by the state to pave roads. Around the year 1300, the geographer Yaqut (1179-1229) described in detail how 'asphalt' was made from this pit: *"There are workers who collect (bitumen) from the spring in woven reed baskets and pour it over the ground. They also have large iron kettles placed over cauldrons which they load with known proportions of bitumen, water and sand. When the stirred mixture reaches the right consistency it is poured over the ground as 'pavement'.*

(OPPOSITE) THE BURNING OF MECCA (AD 1400), PERSIAN MINIATURE.
"…the burnings of Mecca and Medina demonstrate that the Muslim forces had access to oil and were equipped to use it".

COIN SHOWING CONSTANTINE IV.
THE BYZANTINE EMPEROR WHO IN AD 680,
USED OIL WEAPON TECHNOLOGIES LEARNED
FROM A SYRIAN REFUGEE, TO BREAK THE
ARAB SIEGE OF CONSTANTINOPLE. THIS
LED TO THE MISNOMER 'GREEK FIRE'.

People visit this site on outings and to drink the water that comes out with the bitumen. They also bathe in the water, for it is as good in clearing pustules and other diseases as public baths and other remedies." In Europe, roads paved with anything but flagstone and cobbles were unknown until 1838, when asphalt was first laid on a street in Paris.

By the year AD 900, the strategic and economic importance of oil led the Abbasid caliph in Baghdad to appoint what we may today call an 'oil czar' in every major oil producing district. The wali al-naft, as he was called, acted as the eyes, ears and above all the tax-collecting arm of the caliph in the lucrative oil works. Two developments that took place around the year AD 850 increased the power and prestige of the oil czar. The first was the increased demand from a new fighting unit in the Arab regular army called the *'naffatun'* or 'naphtha troops'. The second was the introduction of refined lamp oil, or kerosene, manufactured from crude oil into white naphtha or *naft abyad*, by distillation.

A CRUDE MEDIEVAL ILLUSTRATION SHOWING A NAVAL NAPHTHA CANNON.

BRONZE MACEHEADS WITH BITUMEN SETTINGS,
FROM PALACE OF NIMRUD 850 BC.

GOLD AND ASPHALT BADGES, FROM NORTH WEST IRAN 1350 BC.
THESE MAY HAVE ORIGINALLY BEEN WORN ON CLOTHING, OR USED AS
DECORATIONS FOR HORSE HARNESSES.

GOLD ORNAMENT, FROM THE OXUS TREASURE, NORTHERN BACTRIA 500 BC.
ORIGINALLY INLAID, THIS DEPICTS A MYTHICAL CREATURE WITH THE BODY OF A
WINGED STAG AND THE HEAD OF A HORNED LION. IN SCYTHIAN 'ANIMAL STYLE'
AND SHOWING THE INFLUENCE OF THE NORTHERN STEPPES.

The medieval Arabs used an apparatus called *al-inbiq*, a batch-process still whose name we have taken into English as alembic.

Essentially, the alembic consisted of three parts: a gourd-shaped lower flask called the cucurbit in which the crude oil was heated; a spouted condenser that sat atop the cucurbit and received the vapours that rose from the oil; and a receiver at the end of the condenser's spout for collecting the clear distillate.

In Abbasid times, every school of chemists had its own variation of the alembic. Some were made of blown glass like today's labware, others were made of ceramic, copper or brass. The Syrian naturalist al-Dimashqi, born in 1256, left us a hand-drawn sketch of an industrial-size still he saw in a shop in Damascus and wrote that in his time there was a quarter in the city known as Suq al-Qattarine, the distillers market.

(OPPOSITE) A KURDISH WARRIOR - AMADEO PREZIOSI (1816 - 1882).

The first Muslim scholar to write about the distillation of oil was the Persian-born Muhammad ar-Razi, who spent most of his adult life in the ninth century as a physician and chemist in Baghdad. In his 'Kitab al Asrar,' or 'Book of Secrets', he mentions the use of 'Naffata', or 'Kerosene lamp', for heating and lighting, establishing that such devices were in existence in the Muslim world more than a thousand years before they became known in the West. Distillation made possible the use of kerosene throughout the entire Near East, bringing it to such places as Palestine, Yemen and the Hadhramaut, and Egypt, none of which had any surface deposits of oil to speak of, but all of which had substantial deposits of either oil shale or bitumen. With either of these substances, a reasonably good grade of kerosene could be obtained by first extracting the oil by heating the rock, and then distilling the oil in the alembic. As knowledge of oil grew, so did further refinements of its military applications. With the systematic exploitation of the large pits, enough oil was obtained to burn down both Baghdad and Cairo, two of the region's largest cities, in catastrophes that far surpassed the siege of Mecca. Baghdad in the year AD 800 was the undisputed capital of the Muslim state and the seat of Harun al-Rashid, one of the most powerful rulers of his time. The city's position on the west bank of the Tigris allowed easy communication with the outside world. One-third of the city's area was occupied by the 'Golden Gate' or the royal palace, and the rest of it was home to more than one million inhabitants. As one Arab writer of the time put it *"a city with no peer throughout the whole world"*, but by AD 813, the royal palace was gone and much of the city lay in ruins.

(OPPOSITE AND ABOVE) *"…the Medieval Arabs used an apparatus called al-inbiq, a batch-process still whose name we have taken into English as alembic"* - *still used in modern distillation.*

(LEFT)
GOLD AMULET CASE,
FROM THARROS
700 BC.
THE CASE ORIGINALLY
CONTAINED AN AMULET
WRITTEN ON PAPYRUS
OR GOLD LEAF.

(FAR LEFT)
GOLD VASE
PENDANT, FROM
THARROS 700 BC.

Between AD 809 and 813 Iraq and Persia engaged in a civil war that pitted two of Harun al-Rashid's sons, Amine and Ma'mun, against each other. Ma'mun hoped to trap his brother in Baghdad. He had his 'naphtha troops', equipped with hundreds of mangonels, bombard a section of the city called Harbiyyah with barrels of burning naphtha. The resulting fires eventually engulfed the rest of Baghdad, causing its inhabitants to flee. So total was the destruction that it was not until six years later, in AD 819, that Ma'mun, who had succeeded his father, re-entered the city and began its reconstruction.

Cairo's turn came three centuries later, in the thick of the Crusades. By that time, petroleum-based weapons had reached further levels of sophistication.

In 1167, the crusader king of Jerusalem, Amalric I, decided that a victory in Egypt would provide the resources to resist Syria.

At the head of an army of several thousand, Amalric crossed the Negev and Sinai Deserts and arrived at Bilbeis, North East of Cairo, which he sacked after slaughtering nearly all its inhabitants.

(OPPOSITE PAGE) *"…there was a quarter in the city known as Suq al-Qattarine, the distillers market"*.

BRONZE LAMP, FROM BABYLON 500 BC.
DESIGNED FOR SUSPENSION FROM THREE LOOPS.

ENGRAVED SILVER
LAMP, FROM UR
2600 BC.

He then sent word to the Egyptian caliph 'Athid, only 18 years old, to quit the city or face the fate of Bilbeis. But 'Athid's vizier, Shawar, an old adversary of Amalric swore to deny him the satisfaction of capturing the city intact. He is said to have shouted *"they will only capture a mound of rubble"*.

The horrors of the ensuing days were recorded vividly by the Egyptian historian al-Maqrizi:

"Shawar ordered that Fustat be evacuated. He forced (the citizens) to leave their

POTTERY LAMP, FROM NIMRUD 700 BC.
"…'Naffata', or kerosene lamps, for heating and lighting, were in existence in the Muslim world more than a thousand years before they became known in the West".

money and property behind and flee for their lives with their children. In the panic and chaos of the exodus, the fleeing crowd looked like a massive army of ghosts…

Upon arrival in Cairo, they gathered in the downtown district. Some took refuge in the mosques and bath-houses … awaiting a Christian onslaught similar to the one in Bilbeis. Shawar sent 20 000 naphtha pots and 10 000 lighting bombs ('Mish`al') and distributed them throughout the city.

Flames and smoke engulfed the city and rose to the sky in a terrifying scene. The blaze raged for 54 days…"

That the whole city could be set alight with 'naphtha pots' on relatively short notice is an indication that during the era of the Crusades, oil was readily available in military warehouses and that, in Cairo at least, it was available in large quantities. The oil in Cairo may have been imported from Iraq, Persia or the Caucasus, but most likely it was brought

from the wells of Jabal Tor on the South-Western edge of the Sinai, a seepage which had been exploited since Roman times. Nothing in the writings of Egyptian historians about the burning of Cairo gives any clue as to the use of 'naphtha pots'. It was not until 1916 that two archaeologists, Ali Bey Bahjat, director of the Cairo Museum, and Albert Gabriel, a Frenchman, unravelled a mystery that tells a story of Muslim technology at a time when Islam was threatened by the crusaders and the Mongols.

It was not until ad 1400 that 'Greek fire' began to be replaced by gunpowder. Naphtha had been the primary weapon in the arsenal of many armies in ancient times and had influenced the course of military history.

The two men set out to excavate the rubble of Old Cairo in search of the peculiar broken clay pots, resembling hand grenades, that Egyptian 'night diggers' occasionally sold to Western visitors. It was suspected that these pots were used to burn the city in medieval times. In the 1940's, the pots caught the attention of yet another French scientist, Maurice Mercier. He noticed that those that had the strongest walls and the most aerodynamic designs often had their tops broken off, while the rest of the body was intact. Only a powerful internal explosion, he reasoned, could have caused such clean, sharp fractures. He had a number of the pots carefully examined and discovered that they contained traces of nitrates and sulphur, two essential ingredients in gunpowder. Apparently, the several varieties of 'naphtha pots' used to destroy Old Cairo, were each something between a Molatov cocktail and a crude hand grenade, filled with a volatile jelly of kerosene, nitrates and sulphur.

Clearly, the makers of the firebombs were technicians with a sophisticated knowledge, not only of explosives and incendiaries, but also of soil science and ceramics. Their makers must also have known mechanics and at least the rudiments of aerodynamics.

ISKANDAR'S IRON CAVALRY FROM THE SHAHNAMA (BOOK OF KINGS) BY FIRDAWSI AD 1330.

NAPHTHA POTS, FROM SYRIA OR IRAQ AD 1200.

These discoveries shed further light on a unique Arab manuscript brought to the Bibliothèque Nationale in Paris in the mid 19th century entitled 'The Book of Horsemanship and the Art of War.' Written in 1285 by Najm al-Din Ahdab, a Syrian officer, the book is packed with information on how to distil oil to make kerosene; how to prepare explosives from gunpowder; how to fit the multiple fuse into the various kinds of 'naphtha pots' and even how to build 'flying fire' rockets!

In summary, it can be said that the most important period in the history of combustible oil prior to the invention of the internal combustion engine, unfolded during the flowering of the Muslim civilisation in Medieval times.

(PREVIOUS PAGES) RIVER FORTIFICATIONS IN THE GULF OF AQABA.

"...several varieties of 'naphtha pots' used to destroy Old Cairo, each something between a Molatov cocktail and a crude hand grenade, filled with a volatile jelly of kerosene, nitrates and sulphur".

SILVER IBEX HANDLE, FROM THE OXUS TREASURE, NORTHERN BACTRIA 500 BC.

SHELL INLAYS 2600 BC.
THESE ILLUSTRATE THE STANDARD
SUMERIAN METHOD OF DECORATING
WOODEN ITEMS SUCH AS MUSICAL
INSTRUMENTS AND FURNITURE,
USING A COATING OF BITUMEN
INLAID WITH LAPIS AND OTHER
PRECIOUS STONES.

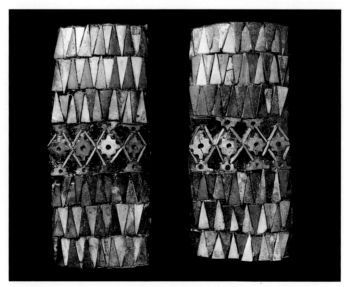

(OPPOSITE PAGE) SILVER RHYTON (HORN-SHAPED DRINKING CUP) IN THE FORM OF A WINGED GRIFFIN 500 BC.

CHAPTER
FOUR

———

THE

SILK
ROAD

FIGURE OF GUANDI (GOD OF WAR), CH'ING DYNASTY AD 1600.

here was a river in the distant lands to the West, said an ancient Chinese sage, *"whose waters though mighty could nevertheless bear not the weight of a single wild-goose feather"*. Yet, added a later sage, *"without the help of a boat made of feathers you will not be able to cross it."*

The name of this river, appropriately enough, was *Jo Shui*, which meant 'weak-water.' and which scholars now tell us, was one the Chinese names for naphtha, at least 2 000 years ago. As in the Near East and Central Asia, oil played an important role in the culture

and economy of ancient China, especially in the three Western provinces of Kansu, Szechuan and Shensi. Ancient records show that in these three regions oil has been exploited since the early Han period around 206 BC.

(PREVIOUS PAGES) *"…oil played an important role in the culture and economy of ancient China"*.

GOLD PLAQUE DECORATED WITH DRAGON DESIGN, MING DYNASTY AD 1400.
THE OPEN GOLDWORK IS DECORATED WITH PRECIOUS STONES, BITUMEN WAS OFTEN USED AS AN ADHESIVE SETTING.
THE PLAQUE MAY HAVE BEEN SOWN ONTO OFFICIAL ROBES AS A BADGE OF RANK.

RED LACQUER BOXES, CH'ING DYNASTY AD 1700.
DECORATED WITH 'DRAGONS IN WAVES' AND CHRYSANTHEMUM DESIGNS.

One of the earliest descriptions of a natural oil seepage in Chinese writing is by the court-scholar Tang Meng in AD 200. In his book Po Wu Chi or 'Record of the Investigation of Things,' dating to about AD 190, he mentioned that in the mountains south of Yenan, in Shensi, there are certain rocks from which springs of 'water' arise.

"They form pools as big as bamboo baskets," he said, " and it flows away in small streams. This liquid is fatty and sticky like the juice of meat. It is viscous like uncongealed grease.

If one sets light to it, it burns with an extremely bright flame. It cannot be eaten". Tang Meng then added that the local people call this greasy substance *shih ch'i* or 'stone lacquer', because they used it, among other things, in the making of lacquers. *Shih ch'i* and *Jo Shui* are only two of the many ancient Chinese names for oil.

Other more common names, included *shih yu*, or 'stone oil', which was common in AD 600 and translates exactly as 'petroleum' in the Greek language, as well as *shih chih shui* (stone fat water), and *menghuoyu* (fierce fire oil).

(OPPOSITE PAGE) SMALL RED LACQUER BOTTLE WITH INTRICATE CARVING, CH'ING DYNASTY AD 1750.

"…in AD 668, the Chinese ambassador and his suite cooked their meal using oil to fuel the fire".

In addition to the oil pools of Shensi, Chinese texts also speak of important oil deposits in Kansu, Szechuan and near the rugged border with Nepal. In fact, in the Tang Buddhist encyclopedia compiled by the monk Tao-Shih in AD 668, there is a delightful story about a Chinese ambassador named Wang Hsuan-Tshe.

On his way to his post on the China-Nepal border, this official took time to visit a nearby oil pool which he described as *"a little lake of 'water' which would burn. If one sets light to it, a brilliant flame would appear all over its surface as if coming forth from the water. If one poured water on it to extinguish it, the water changed to fire and burned."*

(OPPOSITE PAGE) GOLDEN FIGURE OF A SEATED BUDDHA ON A LOTUS BASE, MING DYNASTY AD 1600.

As an example of the utility of this flammable liquid, Tao-Shih added that *"the Chinese ambassador and his suite cooked their meal on it"* - which goes to show that crude oil was known as a cooking fuel even in remote places more than fifteen hundred years ago. A century before the Tang Buddhist encyclopedia was published in AD 500, Chinese texts such as the Shui Ching Chu, or 'Waterways Classic', written by Li Tao Yuan, listed many other uses of oil. These included the making of candles and, amazingly enough, the greasing of moving parts in machinery, notably cart-axles and the bearings of water powered machinery.

Shen Kua, a Chinese inventor writing in about AD 1050, explains how the workers in Fu and Yen, in the provinces of Shensi and Kansu, collected the oil:

FOR THE PAST 2000 YEARS, CHINESE PAINTERS AND CALLIGRAPHERS HAVE USED A PLIABLE BRUSH TO APPLY INKS TO SILK OR PAPER. INK SLABS HAD A FLAT AREA FOR GRINDING THE INK STICK OR CAKE, AND A DEPRESSION FOR HOLDING THE EXCESS FLUID.

PENBOX, MING DYNASTY AD 1520. PAINTED IN COBALT UNDERGLAZE WITH SILVER LIDS.

LACQUERED WOOD WRITING BRUSH,
MING DYNASTY AD 1700.
INLAID WITH MOTHER-OF-PEARL.

"It comes out mixed with water, sand and stones. In the spring the local people collect it with pheasant-tail brushes, and put it into pots where it looks like lacquer."
The smoke from the burning of oil fired Shen Kua's imagination as an inventor. In his times, black lacquer for dying clothes and making ink came exclusively from the burning of pine-

wood, which required the cutting of trees. Shen Kua was exceedingly worried about the deforestation of his country: *"Pine forests in Ch'i and Lu have already become sparse. This is now happening in the Thai-Hang mountains. All the woods south of the Yangtze and west of the capital are going to disappear in time if this goes on…"*

Lamenting this fact, this remarkably perceptive man - perhaps the first environmentalist in recorded history - proposes an ingenious solution. Instead of cutting trees and losing the forest, why not use petroleum for the process? *"The petroleum is abundant,"* he noted, *" and more will be formed in the earth, while supplies of pine-wood may be exhausted."* He concluded enthusiastically, *"the ink-makers do not yet know the benefit of the petroleum smoke."*

While the accounts of Shen Kua, Tang Meng and others are astounding in their own right, an even more impressive glimpse of China's past is provided by its unique accomplishments in drilling technology. In this field China truly stood alone.

CHINESE SCHOLARS WERE HELD IN VERY HIGH ESTEEM AND IN ORDER TO BECOME OFFICIALS WERE EXPECTED TO PURSUE LITERARY ACTIVITIES, CULTIVATE THE ARTS OF CALLIGRAPHY AND PAINTING AND BE ABLE TO WRITE IN SEVERAL SCRIPTS.

IN BOTH ANCIENT CHINA AND THE MIDDLE EAST BITUMEN WAS BURNT UNDER CROP TREES TO
PREVENT INFESTATION BY LOCUSTS, CATERPILLARS AND OTHER HARMFUL INSECTS.

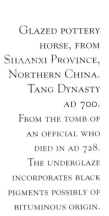

GLAZED POTTERY
HORSE, FROM
SHAANXI PROVINCE,
NORTHERN CHINA.
TANG DYNASTY
AD 700.
FROM THE TOMB OF
AN OFFICIAL WHO
DIED IN AD 728.
THE UNDERGLAZE
INCORPORATES BLACK
PIGMENTS POSSIBLY OF
BITUMINOUS ORIGIN.

HEAD OF A FELINE.
MING DYNASTY AD 1600.
AGATE WITH EYES INSET
WITH BRILLIANTS.

While the Near Eastern and Central Asian people relied exclusively on natural surface seepages for their oil, the ancient Chinese went one step further.

They evidently used deep-drilling to secure not just oil, but natural gas as well. British scholar Joseph Needham - acknowledged expert on the history of technology in ancient China - believes that record depths in excess of 3 000 ft may have been reached in the two provinces of Kansu and Szechuan as early as AD 200. In most cases the wells were sunk to extract brine for the salt industry, but more often than not, and precisely because of the vast underground oil deposits in these two provinces - which are to this day the largest oil exporting provinces of China - the ancient drillers would encounter their salty water not in its 'pure' state, but together with oil and gas.

To tap oil and brine from depths often exceeding 1 000 ft with nothing but the most primitive of tools may sound incredulous at first. Yet there can be no doubt from extensive Chinese literature on drilling that such an engineering feat was commonplace in Szechuan and Kansu around 300 BC.

THE PROBLEM OF
DEFORESTATION CAUSED BY
THE BURNING OF PINE TREES,
IN THE MANUFACTURE OF INKS
AND LACQUER DYES, WAS
IDENTIFIED BY SCHOLARS OF
ANCIENT CHINA.

The boring tool was made of granite, bronze or iron and grinding through the bedrock was an extremely tedious affair, often lasting for years. Workers would employ a large wooden beam with which to hammer the boring tool into the ground.

Meanwhile, at every strike, a second team would rotate the tool and replace the bit whenever it disintegrated. Oil or brine was withdrawn by lowering a long bamboo tube fitted with a

one-way leather valve that let the liquid through only as the tube travelled downward. This is made perfectly clear in

"…To tap oil and brine from depths often exceeding 1 000 ft with nothing but the most primitive tools may sound incredulous, however such an engineering feat was common practice in Szechuan and Kansu around 300 BC".

a Chinese text from about AD 1060 written by a certain Su Tung-Po:
"They use smaller bamboo tubes which travel up and down in the wells; these cylinders have no fixed bottom, and possess an orifice in the top. Each such cylinder brings up large amounts of brine …"

Although seemingly awkward, this Chinese drilling technique worked well - so well, in fact, that it was used by Chinese immigrant workers some 2 000 years later under the description

"...ALL THE WOODS SOUTH OF THE
YANGTZE AND WEST OF THE CAPITAL ARE
GOING TO DISAPPEAR IN TIME IF THIS
GOES ON... INSTEAD OF CUTTING TREES
AND LOSING THE FOREST, WHY NOT USE
PETROLEUM FOR THE PROCESS?"
  Shen Kua - Chinese inventor AD 1050.

'kick her down' when drilling the first oil

wells in southern California.

Whether obtained from natural seepages

or from drilled wells, it was in warfare

where *meng huo yu*, the 'fierce fire oil,'

or naphtha, was best documented.

Court documents from the Northern

Chou period (AD 561 - 577), report that

oil from near the city of Chiu-quan in

Northern Kansu, was used successfully

against the Turks who were besieging the

city. These events occurred only a short

time before naphtha weapons were

reportedly employed by the Arabs and

Byzantines in the Near East, and pose an

interesting question. Did exchanges in

military oil technology occur between

ancient China and the ancient Near East?

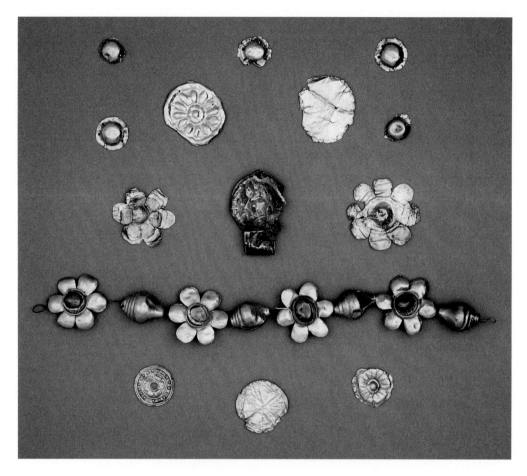

GOLD JEWELLERY OFFERINGS TO BUDDHA AD 200.

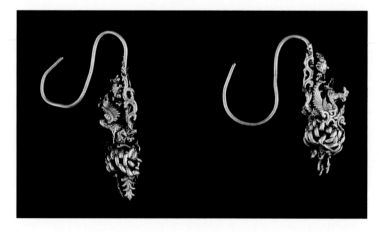

GOLD FILIGREE EARRINGS, MING DYNASTY AD 1500.

SMALL SILVER HORSE AD 1350.
POSSIBLY A BRIDAL DECORATION.

The answer lies in one of the most celebrated transcontinental highways of the ancient world - the 'Silk Road'. By the time Chiu-quan was attacked by the Turks, China had already been trading with the Near East on a massive scale for more than five hundred years. This great commercial highway was formed by a succession of regional roads which created an often tortuous passage between the heartland city of Xian in China and the major cities of Persia, Iraq and Syria. Further important routes led on to Egypt in the South, Baku and Tbilisi to the North, and Istanbul to the West. By 400 BC the Imperial highways of the Persians, including the famous 'Royal Road', had begun to effectively link the lands of the Eastern Mediterranean to the far-flung cities of Central Asia, as far east as Samarkand.

In 100 BC Emperor Wu (141 to 87 BC), expanded his empire westward which marked the formal opening of the 'Silk Road'. China's silk, highly prized by its neighbours, was a major item of trade and in addition camel caravans carried tea, porcelain and jade to the bazaars of the Near East and even to Rome. They returned Eastward with gold and silver.

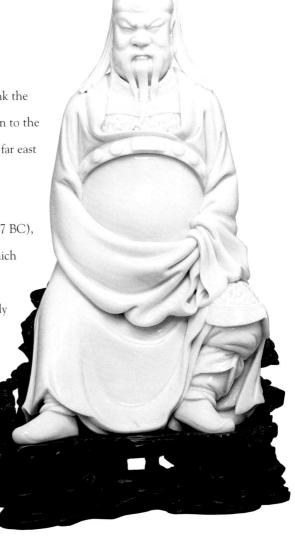

SEATED FIGURE OF GUANDI (GOD OF WAR),
CH'ING DYNASTY AD 1700.

"…there is a delightful story about a Chinese ambassador named Wang Hsuan-Tshe', who, on his way to his post in AD 650, visited a
famous oil pool near the border of China and Nepal: "There is a little lake of 'water' which would burn if one sets light to it…".

175

SMALL BOTTLE WITH ENAMEL DECORATIONS AD 1780.

Remarkably enough, the 'Silk Road' was also the principle channel for oil exports in the ancient world, and in this sense it rightfully qualifies for a second title - the 'Oil Road'. Almost all of the major oil cities and fields of the ancient world lay on the 'Silk Road', from the ill-fated Teflis, burned by the Naphtha Troops of the Caliph in Baghdad, to Baku on the Caspian, to Hit and Baghdad in Iraq, to Susa and Hamadan in Persia, to the oil fields of Turkmenistan, Tashkent and Ferghana. Even Chiu-quan, the Chinese city that defended itself with oil weapons against the Turks, stood on the 'Silk Road'.

The caravans that passed through it laden with silk and silver, no doubt also carried oil to far away markets in the South and North.

With all this traffic, it is not surprising that the 'Silk Road' became the channel through which ancient technologies and oil were exchanged throughout the Asian continent, from the shores of the Mediterranean to the heart of China.

(OPPOSITE PAGE) GREEN JADE VASE, MING DYNASTY AD 1600. ENCRUSTED WITH DARKER JADE, LAPIS AND RUBIES IN GOLD CLOISONNÉ.

CHAPTER FIVE

LAND OF THE NAPHTHA FOUNTAIN

A QUFFA ON THE EUPHRATES - ARTHUR HADDON RBA (1864-1941).
THESE SMALL CRAFT, OFTEN WATERPROOFED WITH BITUMEN, HAVE BEEN USED BY PEOPLE
LIVING ON THE SHORES OF THE EUPHRATES AND TIGRIS FOR THOUSANDS OF YEARS.

*S*ince the collapse of the Soviet Union in 1991, the undeveloped or underdeveloped oil resources of its former Caucasian and Central Asian Republics have been much in the news. However, the existence of rich oil resources in the region from the Western slopes of the Caspian Sea basin to the mountains of Afghanistan has been known for thousands of years.

The ancient Greeks and Romans could not help but notice on their travels the spectacular 'eternal fires' that dotted the landscape all the way from Baku, in present-day Azerbaijan, to Persia and Turkmenistan. Legend has it that one of

the servants of Alexander the Great, accidentally struck oil while trying to pitch a tent for his master by the river Oxus, today's Amu Darya, in Turkmenistan.

According to the Roman historian Pliny the Elder (AD 23 - 79) Alexander the Great himself saw burning oil wells in Bactria - the ancient country that comprised Northern Afghanistan and

parts of the present republics of Uzbekistan and Tajikistan. There, a highly advanced culture prospered from about 600 BC to AD 600. By the time of Pliny, the Greeks and Romans had in fact become aware of practically all the important oil and gas sites from Iraq to the borders of China.

Pliny himself summarises this knowledge very well: "*At Cophantium in Bactria a coil of flame blazes in the night, and the same in Media and in Sittacene, the frontier of Persia: indeed at the White Tower at Susa it does so from fifteen smoke-holes, from the largest even in the daytime.*

(PREVIOUS PAGES)   "*...Al-Mas'udi was so astonished by the amount of oil produced in that town that he called that region 'bilad al-naffata', 'the land of the naphtha fountain'*".

BITUMEN BOWL IN THE SHAPE OF A BISON, FROM SUSA 600 BC.

The Babylonian Plain sends a blaze out of a
sort of fishpool an acre in extent" .

From ancient Media, in North-Western
Iran to Azerbaijan, the eternal fires
probably inspired a religion.
Zoroastrianism, the religion of the
ancient Persians since about 600 BC, was
born in North-Western Iran, practically
amidst the 'Pillars of Fire'. The latter
became centres of worship and
pilgrimage, and even gave their name to
the Zoroastrian priest, the 'Athravan' or
'Keeper of the Fire'.

It may well be that the word Azerbaijan
itself is rooted in the ancient Iranian
'ader-badagan' meaning 'garden of fire'.

BRONZE FIGURINE OF A BEARDED GOD,
FROM NORTHERN MESOPOTAMIA 3000 BC.

Neither Romans nor ancient Persians,
however, left us records of the trade in
oil that must have existed in the region
in ancient times. Such records were only
written centuries later by the Arabs, who
conquered the Caucasus in AD 700.
The newly Muslim lands then included
all the world's known, major oil-producing
regions outside China. In the Caucasus,
the city of Tiflis - now Tbilisi, the capital
of Georgia - grew into a centre of trade
between the Muslim state and Northern
Europe. Gold and silver coins have been
found in the city that date to AD 800
and which were minted in Baghdad,
Muhammadiyyah (in Armenia), Kufa,
Basra, Aran and Balkh, as well as in
Attica and India.

(PREVIOUS PAGE) GOLDEN HEAD OF A YOUTH 500 BC. POSSIBLY PART OF A STATUE OR THE HEAD OF A STAFF.

In addition, Georgia had become an important exporter of naphtha and bitumen to Baghdad. The region was also strategically important to the caliphate, as it was a buffer zone facing Northern Byzantium.

As the Abbasid dynasty weakened after the destruction of Baghdad in AD 813, its North-Western fringes became vulnerable. In AD 843, the Arab Emir of Georgia, Ishaq ibn Isma'il, withheld his annual payment of tribute to Baghdad, and declared his independence from the caliph. To quell the rebellion, Caliph al-Mutawakkil dispatched a punitive expedition led by a Turk named Bugha al-Kabir. He warned Tiflis to surrender or risk a fire the like of which "*exists only in hell*". When the rebels refused, Bugha ordered his *naffatun*, or naphtha troops, to burn the city. So complete was the resultant destruction that it had far-reaching political effects, ending the city's chances of becoming the capital of an Islamic state in the Caucasus. Marco Polo, that extraordinary merchant from the Venetian Republic, was the only European writer in Medieval times to give us a hint of where the oil fields of Georgia might have been.

ROYAL ARCHER OF SUSA 600 BC. THIS GLAZED BRICK RELIEF WAS PART OF A LARGER FRIEZE DEPICTING ROWS OF GUARDS AND APPEARED AT THE ACHAEMENID PALACE AT SUSA, THE CAPITAL OF ELAM, IN THE SOUTH WEST OF ANCIENT IRAN.

183

On his way back from China, around 1291, Marco travelled north from Mosul, in Iraq, into Armenia, to a port city on the Black Sea. Speaking of one of the wonders he encountered there he wrote: *"Bordering upon Armenia, to the South-West, are the districts of Mosul and Maredin … To the North lies Georgia, near the confines of which there is a fountain of oil which discharges so great a quantity as to furnish loadings for many camels.*

The use made of it is not for the purpose of food, but as an unguent for the cure of cutaneous distempers in men and cattle, as *well as other complaints; and it is also good for burning. In the neighbouring country no other is used in their lamps, and people come from distant parts to procure it."*

Today, Georgia has an important oil field that fits Marco's description perfectly: the Patara-Mirzaani field, across the border from Azerbaijan and not very far from the present border with Armenia. Its wells were in fact drilled near ancient oil 'fountains', which lead us to think that one of them may well be that to which Marco Polo refers.

Palmyra - called Tadmor by the Assyrians - is one of the greatest cities of the ancient world. Surrounded by desert, the city was built on an oasis and lies on the ancient trade route midway between the cultivated plain of the Euphrates and the shores of the Mediterranean.

185

Ceremonial mask with gold decoration.

However, scholars in the past have often misinterpreted Marco Polo's words as a reference to the gusher of Baku, whose fame and volume outstripped any other in ancient times.

Built on the Apsheron Peninsula on the Caspian Sea, Baku is today the capital of Azerbaijan, and since the time of Alexander the Great, the city has been known for its fiery wells and coveted for its abundant oil.

As early as AD 642, the Arabs made an attempt on Azerbaijan under an Arabian commander named Mughira ibn Shu'ba, who served also as the governor of Kufa, in Iraq. The region remained under loose Arab rule until the end of the ninth century, with allegiance first to the central government of the Umayyad Dynasty in Damascus and then to the Abbassids in Baghdad. To help him fund the construction of Baghdad, the new capital of the Muslim world, Caliph al-Mansur (AD 754-775) imposed a special 'naphtha tax' on Baku; this marked the first appearance of a state tax on petroleum - a levy with which we are still familiar today.

(LEFT) WOMAN FROM
AZERBAIJAN.
(BELOW) GOLD EARRINGS,
FROM UR 2600 BC.

Baku's oil and the extraordinary sight of its 'eternal fires' caught the eyes of many Arab geographers. Baghdad born al-Mas'udi (AD 956), also known as the Herodotus of the Arabs, was so astonished by the amount of oil produced in the town that he called that region "*bilad al-naffata*", 'the land of the naphtha fountain'. He spent much time exploring and measuring the Caspian Sea, as well as studying the people who lived on its shores.

"*On its shoreline*", he wrote, "*is the place called Baku, which is a naphtha fountain in the kingdom of Shirwan beyond the Golden Hordes and from which white naphtha is obtained.*

There are also burning wells whereby fire emerges from the ground. Across from the naphtha fountain are islands which contain such firewells. These are visible at night from considerable distances."

In the Middle Ages, Baku probably counted only 10 000 to 15 000 inhabitants, many of whom, in one way or another, made their living from the extraction of oil and its transport by ship, cart or barrel-laden camel.

GEORGIAN MAN.

With its excellent harbour, Baku served as the principal export centre for both oil and oil shale from Central Asia to the markets of Persia, the Near East and perhaps, via the Volga River, even parts of Europe. Some two hundred years later, a geographer, Yaqut al-Hamawi, recorded the financial value of Baku's resource in his 'Book of All Lands'. He recorded the value of the oil output from two fields as equivalent to 800g (28 ounces) of gold per day. Such an enormous sum was enough to qualify Baku as one of the wealthiest cities in the Muslim realm.

To the East, across the Caspian from Baku, lay the vast expanse of Central Asia to Tashkent, known as Shash to the Muslims of the Middle Ages. Conquered by the Arab Ziyad ibn Salih in AD 751, Tashkent became the largest and most important city on Islam's Eastern flank, a distinction that it retains even today, as the capital of Uzbekistan. To the Arabs, Tashkent was famous for its fruits and gardens, watered by the Chirchik River.

South of the city however, lay a mountain called Asbara - now called Isfara- which was renowned for its many oil fountains and vast reserves of shale. "*In this area,*" wrote the scholar Qazwini in AD 1200, "*there is a mountain whose rocks are black and burn like charcoal. They are sold by the load at a price of one or two loads per dirham. When this rock burns, it produces an intensely white ash useful in whitening clothes.*"

MOHAMMEDAN FRIAR OF THE ORDER OF THE DANCING DERVISH.
THE SPIRITUAL JOURNEY FOR THE SOUL OF THESE 'SUFI' MYSTICS WAS SEEN AS A PARALLEL TO THE CLEANSING OF IMPURE OIL BY THE ALEMBIC DISTILLER. INDEED THESE PRE-ISLAMIC MONKS WERE AMONG THE FIRST TO EXPLORE ALCHEMY AND THE WONDERS OF THE ALEMBIC.

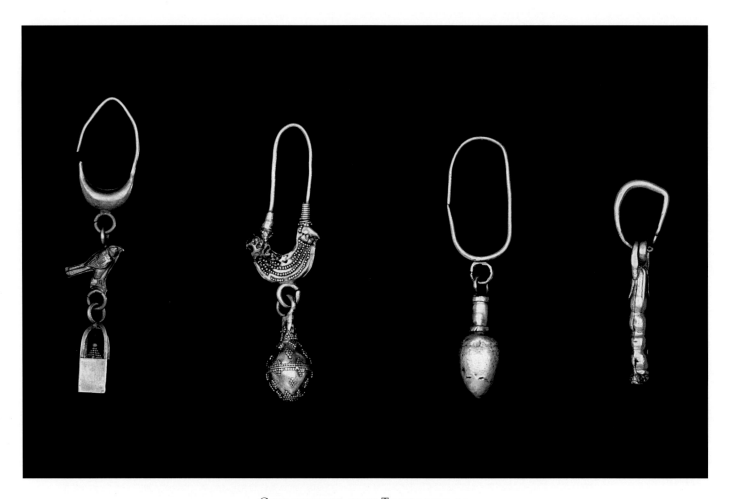

GOLD EARRINGS, FROM THARROS 700 BC.
HAWK WITH BASKET, VASE PENDANT, EGG AND CHRYSALIS DESIGNS.

GOLD NECKLACE, FROM PARTHIA AD 100.
INSET WITH TURQUOISE AND GARNETS.

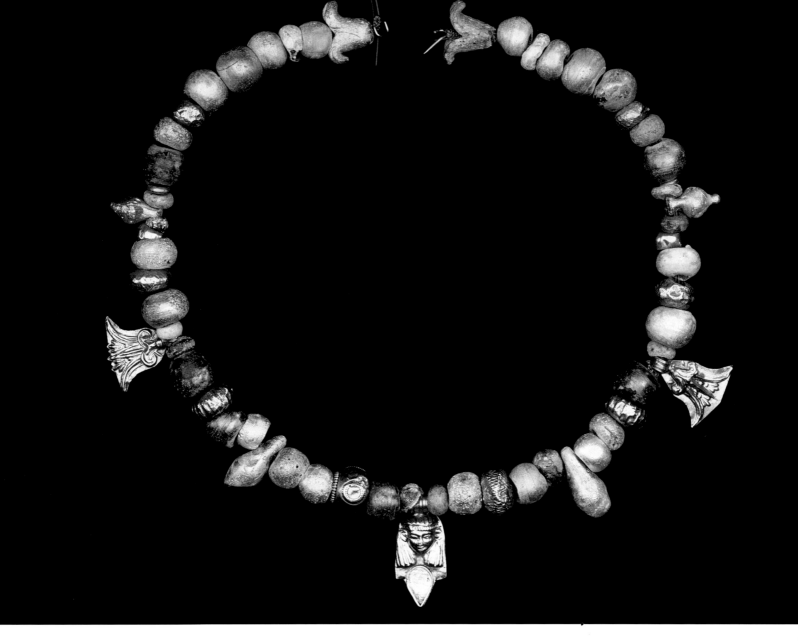

NECKLACE, FROM THARROS 500 BC.
GOLD AND STONE PENDANTS WITH BEADS.

Around AD 850, Ali al-Tabari, a Muslim physician born in the city of Marw in Turkmenistan, wrote a most remarkable book entitled 'Paradise of Wisdom', primarily a book on medicine, but also a hodge-podge of the author's knowledge in such varied fields as physics, history, astrology, botany and horticulture. Ali devoted one chapter to what he called 'resins and fats born of the earth', by which he meant naphtha and bitumen.

Meticulously, he compared the two, their texture, their colour, their uses, and naturally - being a physician - their medicinal values. To be a physician in Ali's time was first and foremost to be an herbalist, and Ali was the quintessential herbalist. Of about a dozen herbal concoctions prepared from naphtha, one in particular stood out as a veritable witches' brew: by combining 35 different herbs and plant extracts, and simmering

them in white naphtha, he obtained a petroleum jelly that, he insisted, could cure almost any ailment: magic spells, poison, snake bites, dog bites, arthritis, gas cramps, pain in the liver, pain in the pancreas and uterus, spasms, colds, chest pain, bronchitis, scabies, kidney stones, eczema, intestinal worms, colitis, skin sores, tuberculosis, loss of memory, incoherence, irregular heartbeats, stuttering, muscle fatigue, chronic fatigue, the list goes on....

"Her people would not answer my queries. Mercy! My liver is like naphtha taken by the flames".
(Basshar ibn Burd AD 783).

GOLD PINHEAD FROM UR 2600 BC.
LAPIS SET ON BITUMEN, LARGE PINS WERE
TYPICALLY USED TO SECURE CLOTHING SUCH
AS CLOAKS.

He also gave the first detailed description of the process of making rose water and other flower extracts in the alembic - the distillation apparatus that was also used to distil crude oil. He was among the first Muslims to write about comets and to describe the fireproof asbestos cloth from Central Asia known as 'salamander skin'. Referring to the legendary dragon, he confidently proposed that the beast played and lived in the fire, only because its skin like the artificial skin of the caliph's naphtha troops, was made of asbestos.

Central Asia, and soon Baku and most of the Near East, fell to the Mongol invasions of AD 1100. Although in 1509, following a period of renaissance in Persia, Baku became a Persian possession called the Baku Khanate. Thereafter the influence of Russia began to grow in the region not for the first time in its history.

Around this time there were frequent wars between the Muslims of the Caspian and the tribes of the Volga called the 'Russ'. One account recalls how the Russ - who were very likely the Scandinavian Vikings who founded Moscow - sailed down the Volga and commenced devastating raids on the Muslim coastal towns.

"...Persian scholar named Abu Tahir al-Fayruzabadi (1329-1414) came to Mecca the holiest city of Islam - to perform the Hajj, the pilgrimage".

"...Baku and most of the Near East, fell to the Mongol invasions".

PERSIAN MINIATURE. GENGHIS KHAN ADDRESSES THE PEOPLE IN THE MOSQUE OF BUKHARA AD 1397.

The Muslims were taken by surprise, having never been attacked from the sea before. The governor of Azerbaijan, Ali ibn Haitham, gathered an army to pursue the enemy to uninhabited islands that faced Baku - the ones with the 'pillars of fire.'

The Russ were forced to retreat and never ventured into the Caspian again.
It was Peter the Great who, in 1700 appears to have been the first European monarch to interest himself in oil, and to foresee the enormous economic potential of petroleum.

In 1723, he gave orders to Matushkin, one of his generals, to take Baku.
Peter wrote *"Of white naphtha send one thousand puds (16 000 kilos or 36 000 lbs), or as much as possible, and find here a refining master."*

AL KIFIL ON THE EUPHRATES,
MESOPOTAMIA - CHARLES CAIN
(1893- 1962).

"…the Arabs prized the oily exudate immensely;
they carried it off like 'plunder of war'".

What is interesting about this request is not so much the size of the shipment but that, first, the Czar used the Arabic term 'white naphtha,' and second that he needed a refining master to accompany the petroleum - an indication that Russia had no refining masters of its own, and needed the technology of the Muslim Caucasus. That this interest on the part of the Czar is documented more than 135 years before the first oil well in the Western world was drilled, is further evidence of the critical importance of Baku in the transfer of oil technology from East to West.

In 1735, the year Empress Anna of Russia restored Baku to Persian rule, an obscure British scientist named Lerche visited the oil fields and recorded his observations. Much of what he described coincided with what the Arabs had reported more than eight centuries earlier. The Englishman toured two large oil fields and wrote that one yielded a crude oil that, when distilled, turned into a bright yellow oil *"resembling a spirit which readily ignited"*.

(PREVIOUS PAGES) RAFT CARRYING THE ASSYRIAN WINGED BULL TO BAGHDAD - F. C. COOPER (1821 - 1878).

BITUMEN BOWL, FROM NINEVEH 3000 BC.

He counted 52 productive wells in one field alone which he said were a great source of wealth for the ruling khan, who owned the wells but leased them to private contractors for a high fee.

Such lease-back arrangements were not new. Cleopatra VII had leased the Dead Sea oil fisheries back to the Nabataeans in 36 BC, and in Baku itself, in an exhausted petroleum pit, a stone bearing

an Arabic inscription has been found that states that the pit had been worked in the year 1597, and gives the name of the operator - Jaz Alia son of Mohammed Nurrs - who leased it.

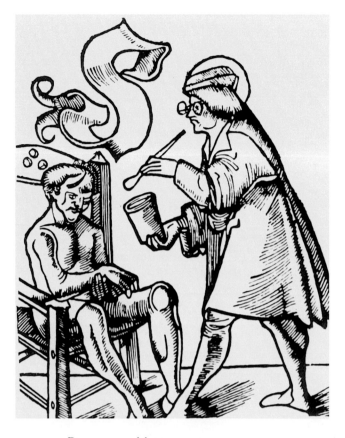

DETAIL OF A MEDIEVAL WOOD CARVING.
*"…Having crushed the roots of plants together with
dried river bitumen, you shall put the preparation on as
a poultice".*

Lerche described how the collected oil was first stored in deep, stone-lined ponds and then carted to Baku in large leather bags. Most residents purchased the oil in its crude state for lighting, he asserted, while others made use of it in their trades. Leather workers especially prized it in the oiling of horse saddles and trappings; cart makers could not do without it as a lubricant for wheel axles, since their alternative was the far more costly whale oil; herbalists used it in the treatment of rheumatics, skin diseases and kidney stones; and veterinary surgeons made it into a 'cattle-dip' to disinfect cattle and other farm animals. As a conclusion to one of his accounts, Lerche gave an interesting theory of his own: because of this miraculous substance, he thought, Baku was spared the ravages of the plague in the Middle Ages.

By Lerche's estimation, the combined output from the two fields he visited was in the neighbourhood of 3 500 tons annually - on the order of 80 or 90 barrels a day.

This is modest production by today's standards, but more than half of it was surplus to Baku's own needs, and thus exportable. It was shipped southward by sea or camel caravan to Persia, or on special Volga River barges to the Northern sides of the Caucasus Mountains.

CHRIST WITH THE CROSS A HELIOGRAPH BY JOSEPH NIEPCE, 1826.
THE HELIOGRAPH PROCESS INVOLVED THE APPLICATION OF A THIN LAYER OF BITUMEN, DISSOLVED IN OIL OF LAVENDER, TO A GLASS OR METAL PLATE.
AN ENGRAVING WAS PLACED OVER THE PLATE AND EXPOSED TO LIGHT FOR SEVERAL HOURS WHICH CREATED A HARDENED BITUMEN PICTURE.
THIS PROCESS WAS USED TO MAKE THE FIRST FIXED PHOTOGRAPH.

Georgien
(Kakhétie.)

A MERCHANT FROM BAKU.
…*"Because of this miraculous substance, Baku was spared the ravages of the Black Plague in the Middle Ages".*

AZERBAIJAN, POSSIBLY DERIVED FROM ADER-BADAGAN MEANING
'GARDEN OF FIRE', HAS BEEN KNOWN FOR ITS SPECTACULAR
'ETERNAL FIRES' SINCE EARLIEST TIMES.

By 1813, however, Persia's hold on Baku had begun to weaken again. After a period of antagonism with Russia, Persia consented to the so-called Gulistan Treaty, under which the Baku Khanate was officially ceded back to the Czar and only Russia was permitted a navy on the Caspian. Inside the khanate however,

Muslim opposition to Russia continued unabated. It came to a head in 1834, when Shaykh Shamil, a local spiritual leader, took to the mountains with his supporters, from where he fought the Russians fiercely for 25 years, until he was captured and killed in 1864.

THE 'PILLARS OF FIRE' AT BAKU AND IN
OTHER PARTS OF AZERBAIJAN SAW THE
BIRTH OF ZOROASTRIANISM - OR FIRE
WORSHIP. A PRACTICE THAT THE ANCIENT
PERSIANS HAD ADOPTED LONG BEFORE.

Oil field near Baku 1860.

This freed Russia's military forces to
further consolidate the Czar's power in
the oil-rich regions East and South of
the Caspian - where Peter the Great had
made his presence felt a century and a
half before - all the way to Tajikistan.
By 1873 the Uzbek Khanate of Kokand,
including the Turkmen nomads and the
cities of Kiva and Bukhara, were
subdued. By 1884, all of present-day
Tajikistan was taken.

ВЪ ПАМЯТЬ ДОБЫЧ
МИЛЛІАРДА ПУДОВЪ СЫРО
ТОВАРИЩЕСТВ
БвъНо
1879

АПШЕРОНСКІЙ
ПОЛУОСТР
БУСАНА
РОМАНЫ
БАЛАХАНЫ
САБУНЧИ
СУРАХАНЫ
КАЛА
БАКУ ЗАВ.БвъНОБЕЛЬ
ПУТА
БИБИЭЙБАТЪ ОС.НАРГИНЪ
ОС.ВУЛЬФЪ
КАСПІЙСКОЕ МОР

(previous pages and above) Fabergé Silver Commemorative Plaque by A. Thielmann, St. Petersburg, 1906.
"In commemoration of the extraction of a million puds of oil by the Company of the Nobel Brothers".
The famous Nobel Brothers Company was established in Baku in 1879 to exploit the oil rich regions of the
Caucasus and to compete against the USA in supplying the growing demand for oil. Ludwig Nobel designed
and built the first ever oil tanker and pioneered the oil industry that we recognise today.

208

The first oil well in North America was drilled in 1858, and by this time Czarist Russia could not contemplate losing Baku which, it was clear to the Czar, was Russia's best hope for countering a potential North American monopoly on the supply of oil. Baku's natural resources attracted the attention of Russian and international concessionaires, especially the Swedish financier and inventor of dynamite Alfred Nobel, who for a while, and because of a lack of modern transportation, resorted to camel caravan for shipping the precious fuel to St. Petersburg and Moscow.

Russia's conquest of these oil rich provinces and their succession under what was later to become the Soviet Union, secured a source of oil which has served until the present decade.

(RIGHT) *"…the gusher of Baku, whose fame and volume outstripped any other in ancient times".*

A B O U T
T H E
A U T H O R

Dr Zayn Bilkadi was born in 1947 in Tunis and now lives in the United States of America.

He obtained a BSc degree in the History of Science from the American University of Beirut, followed by a doctorate in Physical Chemistry from the University of Rochester in New York. He subsequently held academic positions at the University of California in Berkeley and at the University of Minnesota.

Dr Bilkadi is now a Senior Research Specialist at the Corporate Research Laboratories of 3M in St Paul, Minnesota and is currently a holder of nine patents.

Having lectured and written extensively on the History of Science in the Ancient Near East, he is an authority on the subject and his work has been published several times in Aramco World and Chevron World.

This book is his most comprehensive work to date and is written with a deep insight and clear enjoyment of the subject.

LANGUAGE ORIGINS

Development of word-syllabic systems.

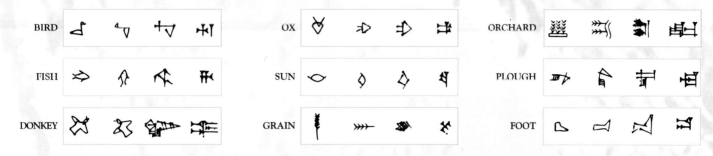

BIRD	OX	ORCHARD
FISH	SUN	PLOUGH
DONKEY	GRAIN	FOOT

Words associated with bitumen are shown to be derived from ancient languages.

MESOPOTAMIAN	IDDU KUPRU **NAPATU** *verb 'to flare up'*		**ESIR**		
	├── AKKADIAN - NORTH MESOPOTAMIA ──┤		SUMERIAN - SOUTH MESOPOTAMIA		
EGYPTIAN	**mrhe.Hr** *cedar oil or calves fat*				
HEBREW	**KOPER** bitumen	**NEFT**	**HEMAR**	**ZEPHET**	
BABYLONIAN	**NAPTU** from the Akkadian verb Napatu				
ARABIC	**KAFR** bitumen	**NAFT** crude oil **HUMAR** may be related to mrhe.Hr	**ZIFT**		**QAYR**
PERSIAN	**NAPTIK**	**MUMIJA** wax		**RHADINACE**	
GREEK	**NAFTA**		**ASPHALTOS** resin **PETRA** rock		
LATIN	**MALTHA**		**ASPHALT**	**OLEUM** oil	**BITUMEN**
ENGLISH	**NAPHTHA**		**ASPHALT**	**PETROLEUM** rock oil *from Petra and Oleum*	**BITUMEN**
CHINESE	**SHIH CHIH SHUI** stone fat water	**JO SHUI** naphtha	**MENGHUOYU** fierce fire oil	**SHIH YU** stone oil *directly translates as petroleum*	**SHIH CH'I** stone lacquer

JEWELLERY

In ancient civilisations, before the advent of the 'money-market', wealth was reflected in the jewellery worn by both men and women - beautiful jewellery of surprisingly elaborate design and intricate detail.

Jewellery was sometimes buried as a sacrificial offering to the gods. It was also interred as part of ancient funerary ceremonies in order to ward off evil spirits and because it was believed that such wealth could be carried to the after-life. Bitumen was sometimes used as an adhesive with which to 'set' gemstones or other inlay such as lapis lazuli, mother of pearl and shell.

AMULET - SECURED
AROUND THE NECK
OR WAIST AS A
CHARM TO WARD
OFF EVIL.
SUCH PENDANTS
SOMETIMES
CONTAINED VERSE
WRITTEN ON PAPYRUS
OR EVEN GOLD LEAF.

ARMLET -
WORN ABOVE
THE ELBOW.

FINGER RINGS - WORN BY PEOPLE
OF MANY DIFFERENT CULTURES
AND VARIED ENORMOUSLY IN
DESIGN. ONE POPULAR DESIGN
WAS THE 'SCARAB' RING - SCARAB
BEING AN ANCIENT GEM CUT IN
THE FORM OF A BEETLE AND
CARVED WITH SYMBOLS.

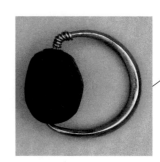

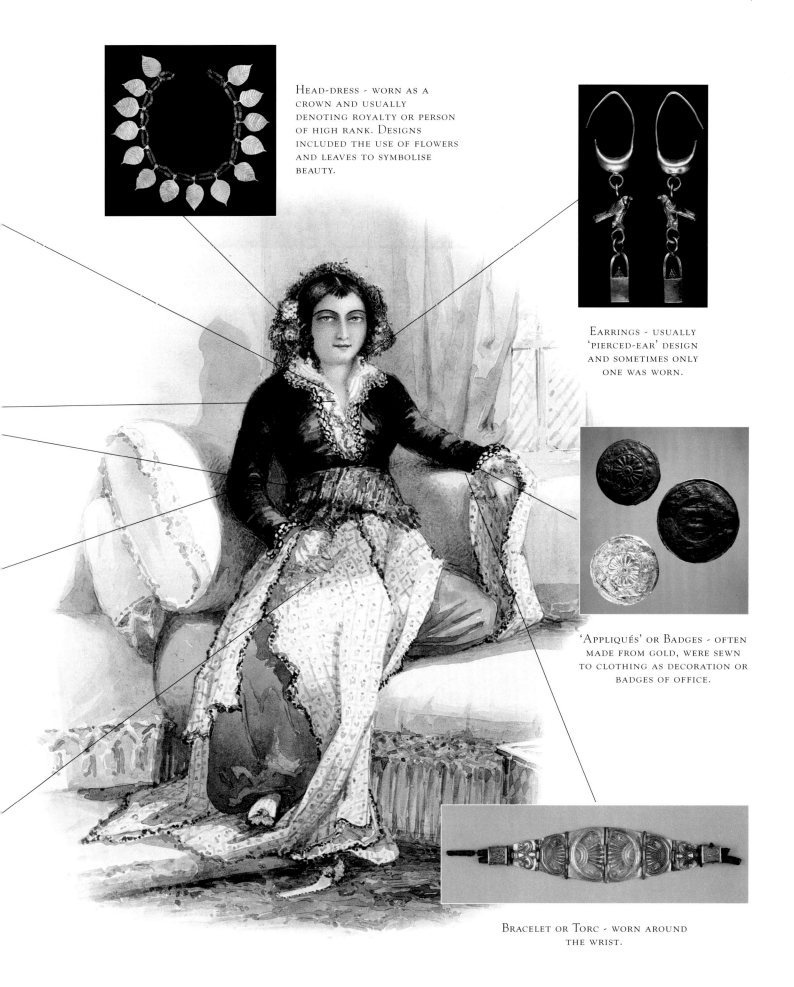

HEAD-DRESS - WORN AS A
CROWN AND USUALLY
DENOTING ROYALTY OR PERSON
OF HIGH RANK. DESIGNS
INCLUDED THE USE OF FLOWERS
AND LEAVES TO SYMBOLISE
BEAUTY.

EARRINGS - USUALLY
'PIERCED-EAR' DESIGN
AND SOMETIMES ONLY
ONE WAS WORN.

'APPLIQUÉS' OR BADGES - OFTEN
MADE FROM GOLD, WERE SEWN
TO CLOTHING AS DECORATION OR
BADGES OF OFFICE.

BRACELET OR TORC - WORN AROUND
THE WRIST.

BIBLIOGRAPHY

BOEDA, ERIC, *Bitumen as a hafting material in Middle Palaeolithic artefacts* in Letters to Nature, VOL. 380, 1996.

BOWERSOCK, GLEN W., *Roman Arabia*, Harvard University Press (1983).

CONNAN, JACQUES; DESCHESNE, ODILE, *Le Bitume à Suse*, Collection du Musée du Louvre (Paris 1996).

DE GENOUILLAC, H., *Inventaire des tablettes de Tello II*, VOL. I (Paris, 1910).

DIODORUS SICULUS, *The Library History*, trans. C.H. Oldfather, The Loeb Classical Library (London and New York, 1933).

FORBES, ROBERT J., *Petroleum and Bitumen in Antiquity*, in Ambix VOL 2 (1938).

FORBES, ROBERT J., *Bitumen and Petroleum in Antiquity*, E.J. Brill (Leiden, 1936).

FORBES, ROBERT J., *Short History of the Art of Distillation*, E.J. Brill (Leiden, 1948).

HENRY, JOHN D., *Baku - An eventful history*, Archibald Constable & Co. (London, 1905).

HERODOTUS, *The Histories*, Penguin Books (New York, 1981).

HITTI, PHILIP K., *History of the Arabs*, St. Martin Press (New York, 1979).

KRAMER, SAMUEL N., *The Sumerians*, University of Chicago Press (Chicago, 1963).

KRAMER, SAMUEL N., *Cradle of Civilization*, Time-Life Books (New York, 1967).

MARCO POLO, *The Travels of Marco Polo*, Ed. M. Komroff, Garden City Publishing (New York, 1930).

AL MAS'UDI, MURUJ AL-DHAHAB WA MA'ADIN AL-GAWHAR, (meadows of gold and mines of gems), *Ed. and trans. de Meynard and de Genouille*, (Paris, 1861-77).

MERCIER, MAURICE, *Le Feu Gregeois, Librairie Orientaliste*, Paul Geuthner (Paris, 1955).

NEEDHAM, JOSEPH, *Science and Civilization in China*, University Press (Cambridge, 1965).

PARTINGTON, JAMES R., *A History of Greek Fire and Gunpowder*, W. Hoffer & Sons, Ltd (Cambridge, 1960).

PLINY THE ELDER, *Natural History*, BOOK II, TRANS. II. Rackham, The Loeb Classical Library (London, 1943).

RUSKA, JULIUS, *Al Razi's Geheimniss Der Geheimniss*, Springer (Berlin, 1937).

SAGGS, H.W.F., *The Greatness that was Babylon*, Sidgwick & Jackson (London, 1966).

SCHEIL, VINCENT, *Les Annales de Tukulti Ninip II*, VOL. V (Paris, 1909).

AL TABARI, MUHAMMAD IBN JARIR, *The History of al Tabari*, Engl. trans. State University of New York Press (Albany, 1985-89).

THUREAUD-DANGIN, FRANCOIS, *Textes Mathématiques Babyloniens*, in Revue d'Assyriologie, VOL. 33 (Paris, 1936).

WALLIS BUDGE, E.A., *Babylonian life and History*, The Religious Tract Society (London, 1925).

ACKNOWLEDGMENTS

The Author wishes to gratefully acknowledge the generosity of Alan Richardson without whose support this book would not have been published.

The Publishers are extremely grateful to the following for their invaluable contributions which have enabled the publication to accurately reflect the diversity of the content.

'Aramco World' Magazine - Robert Arndt, Dick Doughty and members of the editorial staff.

Ashmolean Museum, Oxford University - Dr Roger Moorey.

BP Archive at Warwick University.

BP International.

British Library, London.

British Museum, London - Dr. St. John Simpson, Curator - Department of Western Asiatic Antiquities.

Elf Aquitaine and Fondation Elf.

Harvard University, Massachusetts - Professor A Sabra - Department of History and Science.

Institute of Petroleum, London.

Iran Cultural Heritage Centre, Tehran.

Keeper of the Royal Picture Collection.

Metropolitan Museum, New York.

Musée du Louvre, Paris - Annie Caubet, Conservateur Général chargé - Département des Antiquités Orientales, Odile Deschesne.

Museum of Tehran.

National Gallery, London.

Pelizaeus-Museum, Hildesheim, Germany.

Tina Pennington.

Ahmad Sangari.

Shell International Picture Library, London.

3M.

Bruce Tremayne.

University of Newcastle - Department of Archaeology.

Victoria & Albert Museum, London.

Weizmann Institute of Science - Dr A Nissenbaum.

Stephen Zeal and Claire Nyland for picture research and editing.

Chris Barrett and the Barrett Howe Group for design and print quality.

Illustrations reproduced   © British Museum.

© Musée du Louvre, Départment des Antiquités Orientales.

Courtesy of the Board of Trustees of the V&A.

By permission of Shell International.

By permission of the British Library.

Part of the text and some illustrations in this book were originally published in 'Aramco World' by Aramco Services Company, Houston, Texas. Reprinted by permission.

INDEX

Some dates shown in the book are approximate.

The images in this book have been reproduced using Stochastic Screening Technology.

Printed by The Midas Press, Farnborough, England.

THE ETERNAL FLAME.

Quod non edideris; nescit vox missa reverti - *Horace 65-8 BC.*

You can destroy what you have not yet published.
The word once out can never be recalled.

STANHOPE-SETA
Park Close, Englefield Green, Egham,
Surrey, TW20 0XD England.